ALAN GIBBONS

Department of Computer Science, University of Warwick

Algorithmic graph theory

Published by the Press Syndicate of the University of Cambridge
The Pitt Building, Trumpington Street, Cambridge CB2 1RP
40 West 20th Street, New York, NY 10011-4211, USA
10 Stamford Road, Oakleigh, Melbourne 3166, Australia

First published 1985
Reprinted 1987, 1988, 1989, 1991, 1994

Printed in Great Britain by
Athenæum Press Ltd., Newcastle upon Tyne

Library of Congress catalogue card number: 84-23835

British Library cataloguing in publication data

Gibbons, Alan
Algorithmic graph theory.
1. Graph theory 2. Algorithms
I. Title
511'.5 QA166

ISBN 0 521 28881 9 paperback

Transferred to Digital Reprinting 1999

Printed in the United States of America

UP

To my children
Gabrielle, Chantal and Rosalind,
with love.

Contents

Preface

In the last decade or so work in graph theory has centred on algorithmic interests rather than upon existence or characterisation theorems. This book reflects that change of emphasis and is intended to be an introductory text for undergraduates or for new postgraduate students.

The book is aimed primarily at computer scientists. For them graph theory provides a useful analytical tool and algorithmic interests are bound to be uppermost. The text does, however, contain an element of traditional material and it is quite likely that the needs of a wider audience, including perhaps mathematicians and engineers, will be met. Hopefully, enough of this material has been included to suggest the mathematical richness of the field.

Prerequisites for an understanding of the text have been kept to a minimum. It is essential however to have had some exposure to a high-level, procedural and preferably recursive programming language, to be familiar with elementary set notation and to be at ease with (for example, inductive) theorem proving. Where more advanced concepts are required the text is largely self-contained. This is true, for example, in the use of linear programming and in the proofs of *NP*-completeness.

There is rather more material than would be required for a one-semester course. It is possible to use the text for courses of more or of less difficulty, or to select material as it appeals. For example an elementary course might not include, amongst other material, that on branchings (in chapter 2), minimum-cost flows (in chapter 4), maximum-weight matchings (in chapter 5), postman problems (in chapter 6) and proofs of *NP*-completeness (all of chapter 8). Whatever the choice of material, any course will inevitably reflect the main preoccupation of the text. This is to identify those important problems in graph theory which have an efficient algorithmic solution (that is, those whose time-complexity is polynomial in the problem size) and those which, it is thought, do not. In this endeavour the *most* efficient of the known polynomial time algorithms have not necessarily been described. These algorithms can require explanations that are *too* lengthy and may have difficult proofs of correctness. One such example is graph planarity testing in *linear*-time. It has been thought preferable to go for breadth of material and, where required, to provide references to more difficult and stronger results. Nevertheless, a body of material and quite a few results, which are not easily available elsewhere, have been presented in elementary fashion.

The exercises which appear at the ends of chapters often extend or motivate the material of the text. For this reason outlines of solutions are invariably included. Some benefit can certainly be obtained by reading these sections even if detailed solutions are not sought.

Thanks are due to Valerie Gladman for her cheerful typing of the manuscript. Primary and secondary sources of material are referenced at the ends of chapters. I gratefully acknowledge my debt to the authors of these works. However, I claim sole responsibility for any obscurities and errors that the text may contain.

A. M. Gibbons *Warwick*, January 1984

1

Introducing graphs and algorithmic complexity

In this chapter we introduce the basic language of graph theory and of algorithmic complexity. These mainstreams of interest are brought together in several examples of graph algorithms.

Most problems on graphs require a systematic traversal or search of the graph. The actual method of traversal used can have advantageous structural characteristics which make an efficient solution possible. We illustrate both this and the use of an efficient representation of a graph for computational purposes.

The definitions and concepts outlined here will serve as a foundation for the material of later chapters.

1.1 Introducing graphs

This section introduces the basic vocabulary of graph theory. The subject contains an excess of non-standardised terminology. In the following paragraphs we introduce a relatively small number of widely used definitions which will nevertheless meet our needs with very few later additions.

Geometrically we define a *graph* to be a set of points (*vertices*) in space which are interconnected by a set of lines (*edges*). For a graph G we denote

Fig. 1.1

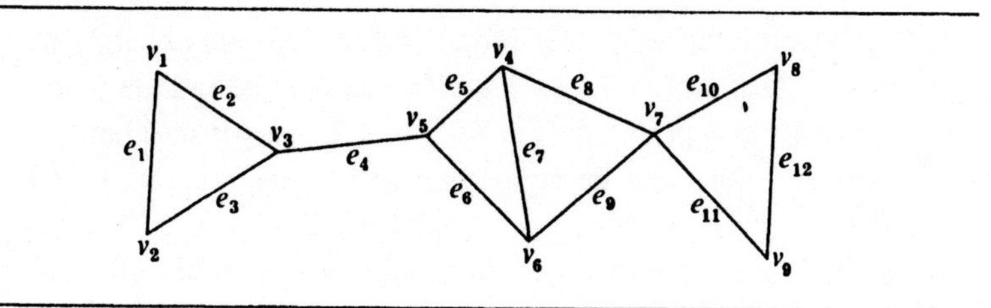

the *vertex-set* by V and the *edge-set* by E and write $G = (V, E)$. Figure 1.1 shows a graph, $G = (\{v_1, v_2, ..., v_9\}, \{e_1, e_2, ..., e_{12}\})$.

We shall denote the number of vertices in a graph by $n = |V|$ and the number of edges by $|E|$. If both n and $|E|$ are finite, as we shall normally presume to be the case, then the graph is said to be *finite*.

We can specify an edge by the two vertices (called its *end-points*) that it connects. If the end-points of e are v_i and v_j then we write $e = (v_i, v_j)$ or $e = (v_j, v_i)$. Thus an equivalent definition of the graph in figure 1.1 is:

$$G = (V, E),\ V = \{v_1, v_2, ..., v_9\}$$

$$E = \{(v_1, v_2), (v_1, v_3), (v_2, v_3), (v_3, v_5), (v_4, v_5), (v_4, v_6), (v_4, v_7), (v_5, v_6), (v_6, v_7), (v_7, v_8), (v_7, v_9), (v_8, v_9)\}$$

If an edge e has v as an end-point, then we say that e is *incident with* v. Also if $(u, v) \in E$ then u is said to be *adjacent to* v. For example, in figure 1.1 the edges e_4, e_5 and e_6 are incident with v_5 which is adjacent to v_3, v_4 and v_6. We also say that two edges are adjacent if they have a common end-point. In figure 1.1, for example, any pair of e_8, e_9, e_{10} and e_{11} are adjacent.

The *degree* of a vertex v, written $d(v)$, is the number of edges incident with v. In figure 1.1 we have $d(v_1) = d(v_2) = d(v_8) = d(v_9) = 2$, $d(v_3) = d(v_4) = d(v_5) = d(v_6) = 3$ and $d(v_7) = 4$. A vertex v for which $d(v) = 0$ is called an *isolated* vertex. Our first theorem is a well-known one concerning the vertex degrees of a graph.

Theorem 1.1. The number of vertices of odd-degree in a finite graph is even.

Proof. If we add up the degrees of all the vertices of a graph then the result must be twice the number of edges. This is because each edge contributes once to the sum for each of its ends. Hence:

$$\sum_i d(v_i) = 2 \cdot |E|$$

The right-hand side of this equation is an even number as is the contribution to the left-hand side from vertices of even-degree. Therefore the sum of the degrees of those vertices of odd-degree is even and the theorem follows. ∎

A *self-loop* is an edge (u, v) for which $u = v$. An example is e_1 in the graph of figure 1.2(a). A *parallel* edge cannot be uniquely identified by specifying its end-points only. In figure 1.2(a), e_2 is parallel to e_3. In this text we shall normally be concerned with *simple* graphs, that is, graphs which contain no self-loops or parallel edges. Of course, every graph has an *underlying* simple graph obtained by the removal of self-loops and parallel edges. Thus figure 1.2(*b*) shows the simple graph underlying figure 1.2(*a*). By the term *multi-graph* we mean a graph with parallel edges but with no self-loops. From now on we shall employ the term *graph* to mean a simple graph unless we explicitly say otherwise.

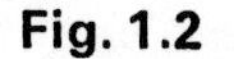

Fig. 1.2

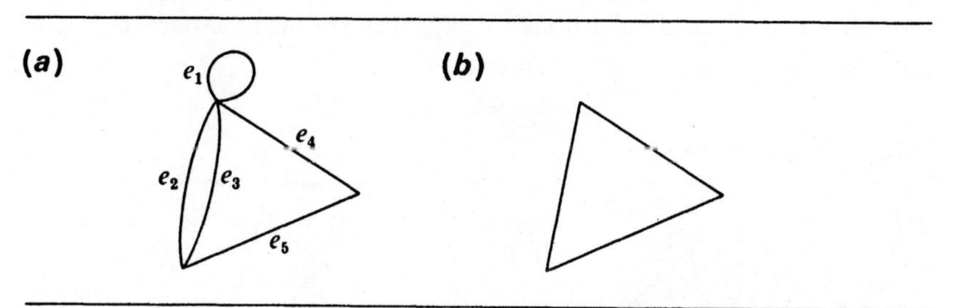

A graph for which every pair of distinct vertices defines an edge is called a *complete* graph. The complete graph with n vertices is denoted by K_n. Figure 1.3 shows K_3 and K_5. In a *regular* graph every vertex has the same degree, if this is k then the graph is called k-*regular*. Notice that K_n is $(n-1)$-regular. Figure 1.4 shows two examples of 3-regular graphs (also called *cubic* graphs) which, as a class, are important in colouring planar maps as we shall see in a later chapter.

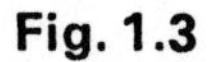

Fig. 1.3

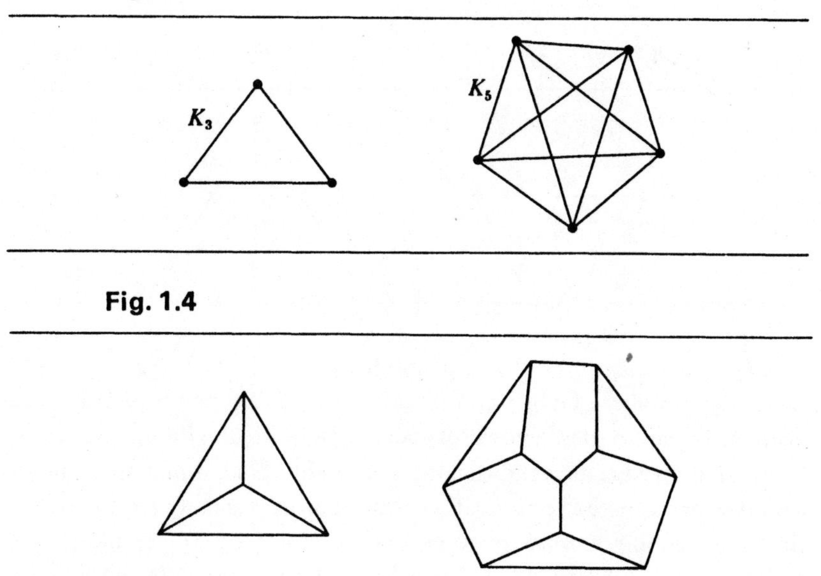

Fig. 1.4

If it is possible to partition the vertices of a graph G into two subsets, V_1 and V_2, such that every edge of G connects a vertex in V_1 to a vertex in V_2 then G is said to be *bipartite*. Figure 1.5(a) and (b) shows two bipartite graphs. If every vertex of V_1 is connected to every vertex of V_2 then G is said to be a *complete bipartite* graph. In this case we denote the graph by $K_{i,j}$ where $|V_1| = i$ and $|V_2| = j$. Figure 1.5(b) shows $K_{2,3}$. There is an obvious generalisation of these definitions for bipartite graphs to k-partite graphs where k is an integer greater than two.

Fig. 1.5

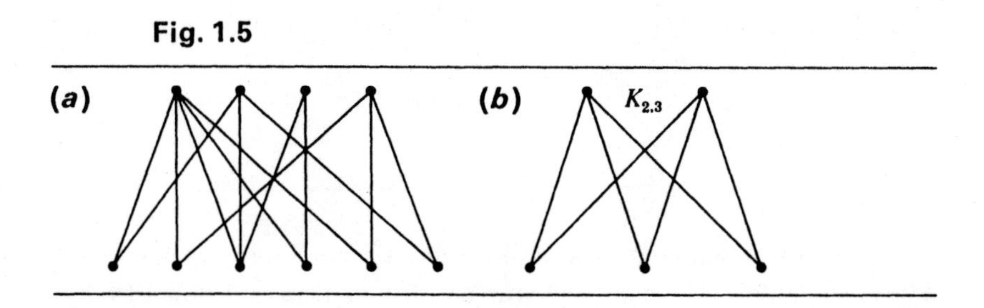

Two graphs G_1 and G_2 are *isomorphic* if there is a one-to-one correspondence between the vertices of G_1 and the vertices of G_2 such that the number of edges joining any two vertices in G_1 is equal to the number of edges joining the corresponding two vertices in G_2. For example, figure 1.6 shows two graphs which are isomorphic, each being a representation of $K_{3,3}$.

Fig. 1.6

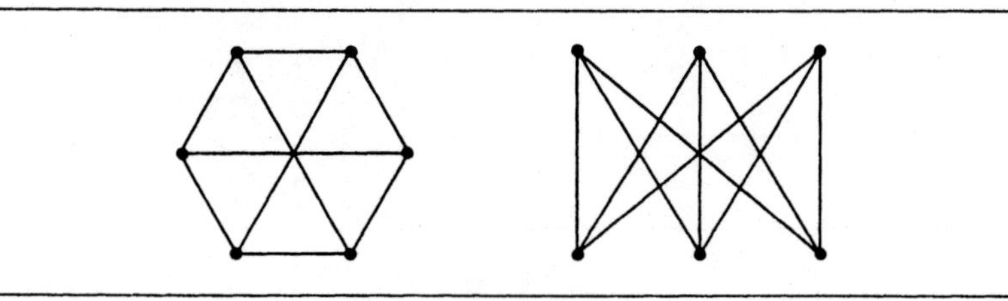

A (proper) *subgraph* of G is a graph obtainable by the removal of a (non-zero) number of edges and/or vertices of G. The removal of a vertex necessarily implies the removal of every edge incident with it, whereas the removal of an edge does not remove a vertex although it may result in one (or even two) isolated vertices. If we remove an edge e or a vertex v from G, then the resulting graphs are respectively denoted by $(G-e)$ and $(G-v)$. If H is a subgraph of G then G is called a *supergraph* of H and we write

$H \subseteq G$. A subgraph of G *induced* by a subset of its vertices, $V' \subset V$, is the graph consisting of V' and those edges of G with both end-points in V'.

A *path* from v_1 to v_i is a sequence $P = v_1, e_1, v_2, e_2, \ldots, e_{i-1}, v_i$ of alternating vertices and edges such that for $1 \leqslant j < i$, e_j is incident with v_j and v_{j+1}. If $v_1 = v_i$ then P is said to be a *cycle* or a *circuit*. In a simple graph a path or a cycle $v_1, e_1, v_2, e_2, \ldots, e_{i-1}, v_i$ can be more simply specified by the sequence of vertices $v_1, v_2, \ldots, v_i$. If in a path each vertex only appears once, then the sequence is called a *simple* path. If each vertex appears once except that $v_1 = v_i$ then P is a *simple* circuit. The *length* of a path or a cycle is the number of edges it contains. Two paths are *edge-disjoint* if they do not have an edge in common.

Two vertices v_i and v_j are *connected* if there is a path from v_i to v_j. By convention, every vertex is connected to itself. Connection is an equivalence relation (see problem 1.9) on the vertex set of a graph which partitions it into subsets $V_1, V_2, \ldots, V_k$. A pair of vertices are connected if and only if they belong to the same subset of the partition. The subgraphs induced in turn by the subsets $V_1, V_2, \ldots, V_k$, are called the *components* of the graph. A *connected* graph has only one component, otherwise it is *disconnected*. Thus the graph of figure 1.1 is connected whilst that of figure 1.9 has two components.

A *spanning* subgraph of a connected graph G is a subgraph of G obtained by removing edges only and such that any pair of vertices remain connected.

Let H be a connected graph or a component. If the removal of a vertex v disconnects H, then v is said to be an *articulation point*. For example, in figure 1.1 v_3, v_5 and v_7 are all articulation points. If H contains no articulation point then H is a *block*, sometimes called a 2-connected graph or component. If H contains an edge e, such that its removal will disconnect H, then e is said to be a *cut-edge*. Thus in figure 1.1 e_4 is a cut-edge. The end-points of a cut-edge are usually articulation points.

A graph with one or more articulation points is also called a *separable* graph. This refers to the fact that the blocks of a separable graph can be identified by disconnecting the graph at each articulation point in turn in such a way that each separated part of the graph retains a copy of the articulation point. For example, figure 1.7 shows the separated parts (or blocks) of the graph depicted in figure 1.1. Clearly, any graph is the union of its blocks.

In some applications it is natural to assign a *direction* to each edge of a graph. Thus in a diagram of the graph each edge is represented by an arrow. A graph augmented in this way is called a *directed* graph or a *digraph*. An example is shown in figure 1.8. If $e = (v_i, v_j)$ is an edge of a digraph then the order of v_i and v_j becomes significant. The edge e is under-

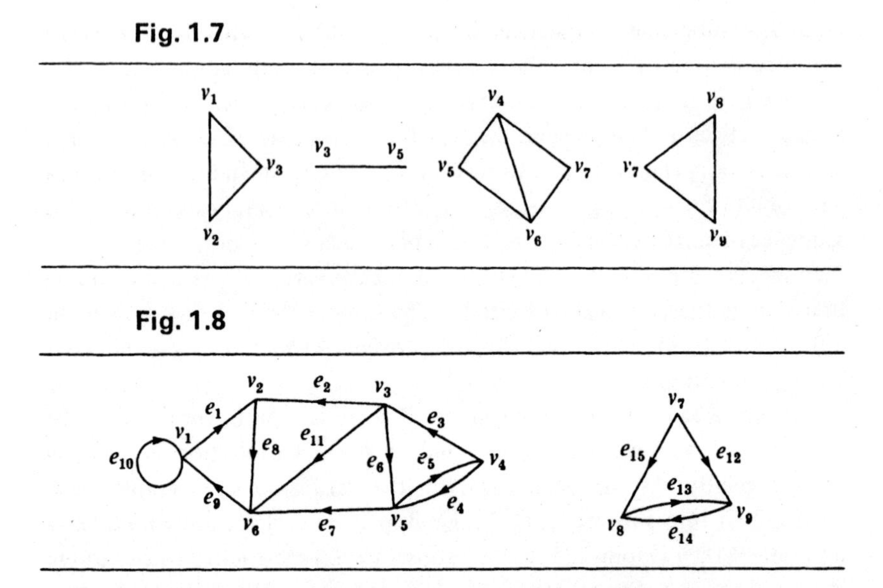

Fig. 1.7

Fig. 1.8

stood to be directed from the first vertex v_i to the second vertex v_j. Thus if a digraph contains the edge (v_i, v_j) then it may or it may not contain the edge (v_j, v_i). The directed edge (v_i, v_j) is said to be *incident from* v_i and *incident to* v_j. For the vertex v, the *out-degree* $d^+(v)$ and the *in-degree* $d^-(v)$ are, respectively, the number of edges incident from v and the number of edges incident to v. A *symmetric* digraph is a digraph in which for every edge (v_i, v_j) there is an edge (v_j, v_i). A digraph is *balanced* if for every vertex v, $d^+(v) = d^-(v)$.

Of course, every digraph has an *underlying* (*undirected simple*) *graph* obtained by deleting the edge directions. Thus figure 1.9 shows this graph for the digraph of figure 1.8. As defined earlier, a path (or circuit) in a corresponding undirected graph is a sequence $S = v_1, e_1, v_2, e_2, \ldots, v_{i-1}, e_i,$ of vertices and edges. In the associated digraph this sequence may be such

Fig. 1.9

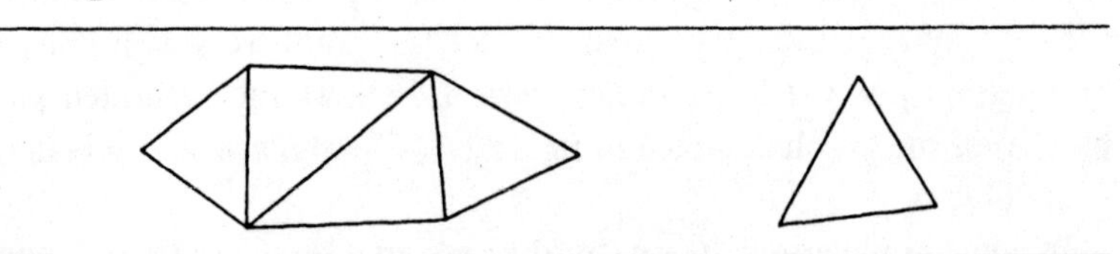

that for all j, $1 \leqslant j < i$, e_j is incident from v_j and incident to v_{j+1}. In this case S is said to be a *directed* path (or circuit). Otherwise it is an *undirected* path (or circuit). Thus in figure 1.8 $(v_2, e_2, v_3, e_6, v_5, e_4, v_4, e_3, v_3, e_{11}, v_6)$ is an undirected non-simple path, while $(v_1, e_1, v_2, e_8, v_6, e_9, v_1)$ is a simple directed circuit. Because in a digraph we can define two different types of paths we can also define two different types of connectedness. Two vertices, v_1 and v_2, are said to be *strongly connected* if there is a directed path from v_1 to v_2 *and* a directed path from v_2 to v_1. If v_1 and v_2 are not strongly connected but are connected in the corresponding undirected graph, then v_1 and v_2 are said to be *weakly connected.*

Both strong connection and weak connection are equivalence relations (see problem 1.9) on the vertex set of a digraph. Of course weak connection partitions the vertices in precisely the same way that connection would partition the vertices of the corresponding undirected graph. Thus for the graph in figure 1.8, weak connection partitions the vertices into the two subsets $\{v_1, v_2, v_3, v_4, v_5, v_6\}$ and $\{v_7, v_8, v_9\}$. The subgraphs induced by these subsets are called the *weakly connected components* of the digraph. On the other hand strong connection partitions the vertices of this graph into the subsets $\{v_1, v_2, v_6\}$, $\{v_3, v_4, v_5\}$, $\{v_7\}$ and $\{v_8, v_9\}$. Each of these subsets induces a *strongly connected component* of the digraph. Notice that each edge of a digraph belongs to some weakly connected component but that it does not necessarily belong to a strongly connected component.

We now briefly introduce an important class of graphs called trees. A *tree* is a connected graph containing no circuits. A *forest* is a graph whose components (one or more in number) are trees. An *out-tree* is a directed tree in which precisely one vertex has zero in-degree. Similarly, an *in-tree* is a directed tree in which precisely one vertex has zero out-degree. A tree in which one vertex, the *root*, is distinguished, is called a *rooted-tree.* In a rooted-tree any vertex of degree one, unless it is the root, is called a *leaf.* As we shall see in theorem 1.2 there is precisely one path between any two vertices of a tree. The *depth* or *level* of a vertex in a rooted-tree is the number of edges in the path from the root to that vertex. If (u, v) is an edge of a rooted-tree such that u lies on the path from the root to v, then u is said to be the *father* of v and v is the *son* of u. An *ancestor* of u is any vertex of the path from u to the root of the tree. A *proper* ancestor of u is any ancestor of u excluding u. Similarly, if u is an ancestor of v, then v is a *descendant* of u. A *proper* descendant of u excludes u. Finally, a *binary* tree is a rooted-tree in which every vertex, unless it is a leaf, has two sons.

Theorem 1.2. If T is a tree with n vertices, then

(*a*) Any two vertices of T are connected by precisely one path.
(*b*) For any edge e, not in T, but connecting two vertices of T, the graph $(T+e)$ contains exactly one circuit.
(*c*) T has $(n-1)$ edges.

Proof. (*a*) T is connected and so there exists at least one path between any two vertices u and v. Suppose that two distinct paths, P_1 and P_2 exist between u and v. Following these paths from u to v, let them first diverge at u' and first converge at v'. That section of P_1 from u' to v' followed by that section of P_2 from v' to u' must form a circuit. By definition, T contains no circuit and so we have a contradiction.

(*b*) Let $e = (u, v)$. According to (*a*) there is precisely one path P from u to v within T. The addition of e therefore creates exactly one circuit $(P+e)$.

(*c*) Proof is by induction on the number of vertices n in T. If $n = 1$ or 2 then, trivially, the number of edges in T is $(n-1)$. We assume that the statement is true for all trees with less than n vertices. Let T have n vertices. There must be a vertex of degree one contained in T, otherwise we could trace a circuit by following any path from vertex to vertex entering each vertex by one edge and leaving by another. If we remove a vertex of degree one, v, from T we neither disconnect T or create a circuit. Hence $(T-v)$ is a tree with $(n-1)$ vertices. By the induction hypothesis $(T-v)$ has $(n-2)$ edges. Hence replacing v provides T with $(n-1)$ edges. ■

We complete our catalogue of definitions by introducing *weighted* graphs. In some applications it is natural to assign a number to each edge of a graph. For any edge e, this number is written $w(e)$ and is called its *weight*. Naturally the graph in question is called a *weighted graph*. The *weight* of a (*sub*)*graph* is equal to the sum of the weights of its edges. Often of interest here is a path (or cycle) in which case it may be appropriate to refer to the *length* rather than the weight of the path (or cycle). This should not be confused with the length of a path (or cycle) in an unweighted graph which we defined earlier.

In the following section we introduce the other central interest of this text, namely, that of algorithmic complexity.

1.2 Introducing algorithmic complexity

Although fairly brief, this introduction to algorithmic efficiency will provide a sufficient basis for all but the final chapter of this text. That chapter provides further insight into what is introduced here, and, in particular, it explores an important class of intractable problems.

Our interest in efficiency is particularly concerned with what is called the *time-complexity* of algorithms. Since the analogous concept of *space-complexity* will be of little interest to us, we can use the term *complexity* in an unambiguous way. The *complexity* of an algorithm is simply the number of computational steps that it takes to transform the input data to the result of a computation. Generally this is a function of the quantity of the input data, commonly called the *problem size.* For graph algorithms the problem size is determined by one or perhaps both of the variables n and $|E|$.

For a problem size s, we denote the complexity of a graph algorithm A by $C_A(s)$, dropping the subscript A when no ambiguity will arise. $C_A(s)$ may vary significantly if algorithm A is applied to structurally different graphs but which are nevertheless of the same size. We therefore need to be more specific in our definition. In this text we take $C_A(s)$ to mean the *worst-case* complexity. Namely, to be the maximum number, over all input sizes s, of computational steps required for the execution of algorithm A. Other definitions can be used. For example, the *expected time*-complexity is the *average*, over all input sizes s, of the number of computational steps required.

The complexities of two algorithms for the same problem will in general differ. Let A_1 and A_2 be two such algorithms and suppose that $C_{A_1}(n) = \frac{1}{2}n^2$ and that $C_{A_2}(n) = 5n$. Then A_2 is faster than A_1 for all problem sizes $n > 10$. In fact whatever had been the (finite and positive) coefficients of n^2 and of n in these expressions, A_2 would be faster than A_1 for all n greater than some value, n_0 say. The reason, of course, is that the asymptotic growth, as the problem size tends to infinity, of n^2 is greater than that of n. The complexity of A_2 is said to be of lower *order* than that of A_1. The idea of the *order* of a function is important in complexity theory and we now need to define and to further illustrate it.

Given two functions F and G whose domain is the natural numbers, we say that the order of F is lower than or equal to the order of G provided that:

$$F(n) \leqslant K \cdot G(n)$$

for all $n > n_0$, where K and n_0 are two positive constants. If the order of F is lower than or is equal to the order of G then we write $F = O(G)$ or we say that F is $O(G)$. F and G are of the *same* order provided that $F = O(G)$ *and* that $G = O(F)$. It is occasionally convenient to write $\theta(G)$ to specify the set of all functions which are of the same order as G. Although $\theta(G)$ is defined to be a set, we conventionally write $F = \theta(G)$ to mean $F \in \theta(G)$. Illustrating these definitions, we see that $5n$ is $O(\frac{1}{2}n^2)$ but that $5n \neq \theta(\frac{1}{2}n^2)$ because $\frac{1}{2}n^2$ is not $O(5n)$. Note also that low order terms of a function can be ignored in determining the overall order. Thus the polynomial $(3n^3+6n^2+n+6)$ is $O(3n^3)$. It is obviously convenient when specifying the order of a function to describe it in terms of the simplest representative function. Thus $(3n^3+6n^2)$ is $O(n^3)$ and $\frac{1}{2}n^2$ is $O(n^2)$.

When comparing two functions in terms of order, it is often convenient to take the following alternative definition. Letting $\lim_{n\to\infty} F(n)/G(n) = L$, we see that:

(i) If $L =$ a finite positive constant, then $F = \theta(G)$.
(ii) If $L = 0$, then F is of lower order than G.
(iii) If $L = \infty$, then G is of lower order than F.

We provide four illustrations:

(*a*) $F(n) = 3n^2-4n+2$ and $G(n) = \frac{1}{2}n^2$. Then $L = 6$, so that $F = \theta(G)$.
(*b*) $F(n) = \log_2 n$ and $G(n) = n$. Then:

$$L = \lim_{n\to\infty} \frac{\ln n}{n}\cdot\log_2 e = \lim_{n\to\infty}\left(\frac{\log_2 e}{n}\right) = 0$$

Here we have used L'Hôpital's rule which states that if

$$\lim_{n\to\infty} F(n) = \lim_{n\to\infty} G(n) = \infty$$

and provided the derivatives F' and G' and the limits exist, then:

$$\lim_{n\to\infty} \frac{F(n)}{G(n)} = \lim_{n\to\infty} \frac{F'(n)}{G'(n)}$$

Since $L = 0$, we see that $\log_2 n$ is of lower order than n.

(*c*) $F(n) = x^n$ and $G(n) = n^k$, where x and k are arbitrary but fixed constants, both greater than one. We define $U(n) = F(n)/G(n)$, so that:

$$U(n+1)/U(n) = x(n/(n+1))^k$$

Thus for fixed k, we can always find a sufficiently large value of n, n_0, say, such that for $n > n_0$:

$$U(n+1) \simeq x\cdot U(n)$$

Hence for $n \geqslant n_0$

$$U(n) \simeq x^{n-n_0}U(n_0)$$

and

$$L = \lim_{n\to\infty} U(n) = \infty$$

So that F, which is exponential in n, is of greater order than *any* polynomial in n.

(*d*) If we take F and G to be as defined in (*c*), and if $H(n) = n!$, then using the same approach as in (*c*) the reader may readily verify that H is of greater order than both F *and* G. In other words, factorial n is of greater order than polynomial n. Moreover, it is of greater order than exponential n.

The order of $C_A(s)$ describes the asymptotic behaviour of $C_A(s)$ as $s\to\infty$. If $C_A(s)$ is $O(F)$, then A is said to be an $O(F)$-algorithm. The *asymptotic complexity* essentially determines the largest problem size that can be handled. If two algorithms for the same problem are of the same order, then roughly speaking, neither performs significantly better than the other. For sufficiently large s, the difference is negligible compared with what it would be for two algorithms of different order.

In table 1.1 we have tabulated various commonly occurring complexities for a range of problem sizes. This table also provides information from which the tabulated numbers of computational steps might be realistically related to computation times.

Table 1.1. *Computation times for a variety of time-complexities over a range of problem sizes*

Time-complexities \ Problem size n	2	8	128	1024
n	2	2^3	2^7	2^{10}
$n \log_2 n$	2	3×2^3	7×2^7	10×2^{10}
n^2	2^2	2^6	2^{14}	2^{20}
n^3	2^3	2^9	2^{21}	2^{30}
2^n	2^2	2^8	2^{128}	2^{1024}
$\sim n!$	2	5×2^{13}	5×2^{714}	7×2^{8766}

$$\begin{aligned} 2^{10} \text{ steps/second} &\simeq 0.9 \times 2^{16} \text{ steps/minute} \\ &\simeq 0.9 \times 2^{22} \text{ steps/hour} \\ &\simeq 1.3 \times 2^{26} \text{ steps/day} \\ &\simeq 0.9 \times 2^{35} \text{ steps/year} \\ &\simeq 0.7 \times 2^{42} \text{ steps/century} \end{aligned}$$

The complexity of an algorithm is important to the computer scientist. One reason for this is that the existence of an algorithm does not guarantee in practical terms that the problem can be solved. The algorithm may be so inefficient that, even with computation speeds vastly increased over those of the present day, it would not be possible to obtain a result within a useful period of time. We need then to characterise those algorithms which are efficient enough to make their implementation useful so that they can be distinguished from those which may have to be disregarded for practical purposes. Fortunately, computer scientists have been able to make use of a rather simple characterising distinction which, for most occasions, satisfies the need. The yardstick is that any O(P)-algorithm, where P is a polynomial in the problem size, is an *efficient* algorithm. Many algorithms have complexities which are exponential, or even factorial, in the problem size. From our illustrations of determining the relative order of functions, we see that these algorithms cannot, at least according to the technical definition, be regarded as efficient. In table 1.1 we can see, for the examples shown, that there is a phenomenal difference in the growth of computation times for these algorithms compared with that for some *polynomial time* algorithms. (Note that $n \log_2 n$ is $O(n^2)$.) We must, however, not lose sight of the dictionary definition of *efficient*. It is quite possible for a particular algorithm to be inefficient in the technical sense and yet to be preferable in practice.

The technical distinction, as it has been drawn, between efficient and inefficient algorithms can be a crude one since it takes no account of the coefficients or the degree of a specific polynomial in question. For very small problem sizes, we can see from the table 1.1 that algorithms with complexity 2^n or $n!$ are actually more efficient than an algorithm of complexity n^3. The range of this greater efficiency would clearly extend to much greater problem sizes if the comparison had been made with a complexity of $1000n^3$ or with a complexity of n^{100}, say. However, it is true that, in practice, these considerations are uncommonly an issue because the polynomials encountered are usually of low degree and contain modest coefficients. In a different vein, we must also remember that the complexity of an algorithm describes its worst case behaviour. Its average behaviour may be a much more attractive prospect. A well-known example, which falls outside our technical definition of efficiency, concerns the problem of *linear programming*. Here, as we point out at the end of the appendix on linear programming, there is a commonly used exponential-time algorithm which is nevertheless efficient in practice and yet there exists a polynomial time algorithm which is at present hopeless from a practical standpoint.

It might be thought that our specification of efficiency will lose its usefulness as new generations of computers operate at higher and higher speeds. It is perhaps remarkable that this is not the case. We can best see this by tabulating the maximum problem sizes that can be solved with various time-complexities, over a common time period, as the speed of computation is increased. This has been done in table 1.2. This demonstrates that higher computation speeds have a significant multiplicative effect upon the maximum problem size that can be solved by polynomial time algorithms but only a marginal additive effect for exponential-time algorithms. Of course, this only serves to enhance our notion of what algorithms may be regarded as efficient.

Notwithstanding our earlier caveats, we call any problem for which no polynomial time algorithm is known, and for which it is conjectured that no such algorithm exists, an *intractable* problem.

As illustrations, we now analyse two algorithms. The first is a well-

Table 1.2. *The effect of faster computation speeds on the largest problem size solvable in a given time by some polynomial and exponential-time algorithms*

Time-complexity	At present speeds	$2^3 \times$ faster	$2^7 \times$ faster	$2^{10} \times$ faster
n	N_1	$8N_1$	$128N_1$	$1024N_1$
n^2	N_2	$2.8N_2$	$11.3N_2$	$32N_2$
2^n	N_3	N_3+3	N_3+7	N_3+10
8^n	N_4	N_4+1	$N_4+2.3$	$N_4+3.3$

known one due to Dijkstra which finds the shortest path from a specified vertex in a weighted graph to any other vertex, or indeed to *all* other vertices. The second algorithm solves the seemingly similar problem of finding the *maximum* length simple path between any two specified vertices in a similar graph. Both algorithms work for directed or for undirected graphs.

For the purpose of communicating algorithms, we assume that the reader has some experience of computer programming in a high-level language such as ALGOL or PASCAL. In this text we describe algorithms in terms of a simple model language which will require no formal definition for the experienced programmer. Also our programs concentrate on what is basically algorithmic and avoid any inessential verbosity and unbending syntax than an actual programming language might force upon us.

Dijkstra's algorithm is shown in figure 1.10. For an undirected graph, we replace each edge (u, v) by two directed edges (u, v) and (v, u). Each vertex v of a graph $G = (V, E)$ which is subjected to the algorithm, has an associated label $L(v)$. This is initially assigned a value equal to the weight $w((u, v))$ of the edge (u, v), where u is the vertex from which path lengths are to be measured. If u and v are distinct and $(u, v) \notin E$ then $w((u, v)) = \infty$, while $w((v, v)) = 0$. On termination of the algorithm $L(v)$, for all $v \in E$, is the length of the shortest path from u to v. The algorithm works by constructing a set $T \subseteq V$ in such a way that the current shortest path from u to any $v \in T$ only passes through vertices in T. Figure 1.11 illustrates an application of Dijkstra's algorithm. For the graph shown there the table lists the values of the $L(v)$ and T for each iteration of the while-statement of line 4 of figure 1.10.

Before establishing the complexity of Dijkstra's algorithm we prove in theorem 1.3 that it does indeed do what is claimed for it.

Theorem 1.3. Dijkstra's algorithm finds the shortest path from u to every other vertex.

Proof. We first prove by induction on the size of T that:

(*a*) for every $v \in T$, $L(v)$ is equal to the length of the shortest path from u to v, and

(*b*) for every $v \notin T$, $L(v)$ is equal to the length of the shortest path from u to v which, apart from v, passes only through vertices in T.

As the basis of our induction notice that when $|T| = 1$ at line 3 of figure 1.10, lines 1 and 2 have initialised the labels to satisfy (*a*) and (*b*).

The inductive step, embodied in line 6, adds to T the vertex v' which has the smallest label of those not yet in T. By the inductive hypothesis, just before v' is added to T, $L(v')$ is equal to the shortest path from u to v' which, apart from v', only utilises vertices in T. Suppose that when v' is added to T, $L(v')$ is *not* equal to the shortest path from u to v'. Then a shortest path must contain, apart from v', at least one other vertex not in T. Let v'' be the first such vertex in tracing this path from u. Then the distance along this path to v'' (which lies entirely within T and which is the shortest from u to v'' – otherwise an even shorter path from u to v' would exist) is less than $L(v')$. By the induction hypothesis $L(v'')$ is the distance along this path and so $L(v') > L(v'')$ when v' was added to T. This contradicts line 5 of the program and so we conclude that there is in fact no path shorter than $L(v')$ when v' is added to T. Thus (*a*) is maintained as T is added to and so is (*b*) through the statement beginning at line 7.

On completion of Dijkstra's algorithm every vertex is in T and so the theorem follows. ■

Fig. 1.10. Dijkstra's shortest path algorithm.

```
1.  for all v ≠ u L(v) ← w((u, v))
2.  L(u) ← 0
3.  T ← {u}
4.  while T ≠ V do
          begin
5.        find a v' ∉ T such that for all v ∉ T L(v') ≤ L(v)
6.        T ← T ∪ {v'}
7.        for all v ∉ T
              L(v) ← if L(v) > L(v') + w((v', v))
                        then L(v') + w((v', v))
          end.
```

It is easy to see that Dijkstra's algorithm can be implemented so as to run in $O(n^2)$-time. The determination of the minimum $L(v')$ in line 5 can be achieved with $O(n)$ comparisons and line 7 requires not more than n assignments. Both lines 5 and 7 are contained within the body of the while statement beginning at line 4 and this body is executed $(n-1)$ times. The for statement can therefore be made to run in $O(n^2)$-time. The remainder of the program, lines 1 to 3, requires only $O(n)$-time. In exercise 1.16(*b*) we show how the algorithm may be implemented using a *priority queue* and *adjacency lists* both of which we define later. In this form we obtain $O((|E|+n)\log n)$ complexity. For large sparse graphs (i.e., with relatively few edges), this represents a much improved running time.

Fig. 1.11

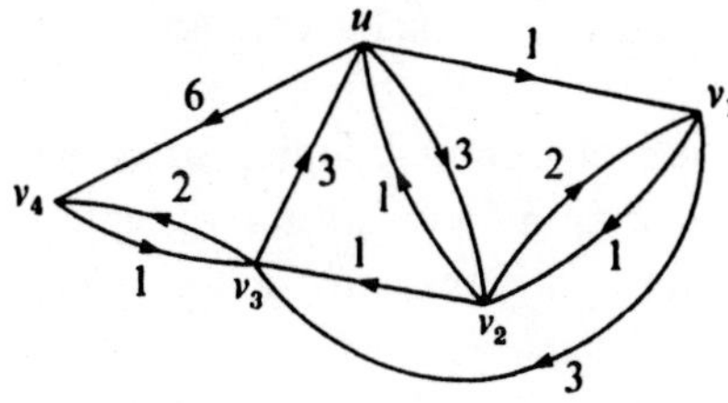

Iteration	v'	$L(u)$	$L(v_1)$	$L(v_2)$	$L(v_3)$	$L(v_4)$	T
0	—	0	1	3	∞	6	$\{u\}$
1	v_1	0	1	2	4	6	$\{u, v_1\}$
2	v_2	0	1	2	3	6	$\{u, v_1, v_2\}$
3	v_3	0	1	2	3	5	$\{u, v_1, v_2, v_3\}$
4	v_4	0	1	2	3	5	V

Dijkstra's algorithm determines the shortest path from u to every other vertex of the graph. If we are simply interested in finding the shortest distance from u to another specified vertex t, then the while-statement beginning at line 4 could be terminated as soon as T includes t. Of course, this would not affect the *order* of the complexity of the computation.

We turn our attention now to the second example. As we stated earlier, this is to find the *maximum*-length simple path between two specified vertices, u and t, of a graph. Any simple path between u and t consists of a subset of the edges of the graph. The algorithm outlined in figure 1.12 enumerates all subsets of E in turn and, for those which represent such a path, a current record of the longest path is kept. It is easy to check that E' is a simple path from u to t in polynomial time. This check is executed for every iteration of the for statement in the algorithm. However, there are $2^{|E|}$ such iterations (because there are $2^{|E|}$ subsets of E) and so without any further detailing of the algorithm, we can see that it is inefficient.

Fig. 1.12. A longest simple path algorithm.

1. $MAXP \leftarrow 0$
2. **for** all subsets $E' \subseteq E$ **do**
3. **if** E' is a simple path from u to t
   then $MAXP \leftarrow$ **if** $w(E') > MAXP$ **then** $w(E')$

We have now seen, by the criteria specified earlier, one algorithm which is efficient and one which is inefficient. As far as the second problem is concerned, we could marginally improve on the complexity of the algorithm supplied by the use of a more cunning or direct enumeration of paths. However, no enumeration for an arbitrary graph is polynomially bounded. In fact, no algorithm is known for this problem which operates within polynomial time. In chapter 8, we shall see that a related decision problem (Given an integer K, does G have a simple path between two specified vertices of length greater than K?) belongs to a large class of problems called *non-deterministic polynomial time complete* (normally abbreviated to *NP-complete*) which are widely held to be intractable.

It is characteristic of the *NP*-complete problems that known algorithms require an exponentially large number of executions of a polynomial time subtask. For example, in the decision problem just mentioned, it is easy to check the length of a given path in polynomial time, but there are an exponentially large number of these. By definition, any one *NP*-complete problem can be transformed into any other within polynomial time. Thus the discovery of a polynomial time algorithm for one would guarantee that such an algorithm exists for any other. So much fruitless effort has been expended in the search for these algorithms that they are thought now not to exist. There is, however, no proof of this conjecture.

We now suspend discussion of *NP*-completeness until chapter 8. There we provide proof that many of those problems to be met in the intervening chapters and for which we can provide no efficient algorithms do in fact belong to this class of *NP*-complete problems. In the interim we shall rely upon the small insight provided here and will identify the problems as they arise.

1.3 Introducing data structures and depth-first searching

We introduce here elementary representations of graphs for computational purposes. We also describe an efficient method for traversing graphs, called depth-first searching. This section then concludes with detailed descriptions of two algorithms made efficient by utilising this material. Another commonly used method for traversing a graph, called breadth-first searching, is described in exercise 1.14 and made use of in exercise 1.15.

1.3.1. Adjacency matrices and adjacency lists

The data structures introduced here are commonly used to represent graphs. In particular, as we shall see later, the use of adjacency lists can make an important contribution to the efficiency of an algorithm.

An *adjacency matrix* for the graph $G = (V, E)$ is an $n \times n$ matrix A, such that:

$$\begin{aligned} A(i,j) &= 1 \quad \text{if} \quad (i,j) \in E \\ &= 0 \quad \text{otherwise} \end{aligned}$$

If G is an undirected graph then $A(i,j) = A(j,i)$, whilst if G is a digraph then A is generally asymmetric. Figure 1.13 illustrates the two cases. A specification of A clearly requires $O(n^2)$ steps. This eliminates any possibility of $O(|E|)$-algorithms if A represents a sparse graph, that is one for which the ratio $|E|/n$ is low. However, as we shall see, $O(|E|)$-algorithms are certainly possible in some cases by making use of *adjacency lists.*

In an adjacency list representation of a graph, each vertex has an associated list of its adjacent vertices. Examples are shown in Figure 1.13. These lists can be embodied in a table T, examples of which are also shown in the diagram. In order to trace the list for v_i, say, in the table, we consult $T(i, 2)$ which points to $T(T(i, 2), 1)$ where the first vertex adjacent to v_i is recorded. Then $T(T(i, 2), 2)$ points to $T(T(T(i, 2), 2), 1)$ where the second vertex adjacent to v_i is recorded, and so on. The list for v_i terminates when a zero pointer is found. Notice the convention of numerically ordering the vertices adjacent to v_i within v_i's adjacency list; this is relevant to understanding some later examples of applying algorithms. Clearly, T has $(n+|E|)$ rows for a directed graph and $(n+2|E|)$ for an undirected graph. In some circumstances it is additionally useful to use doubly linked lists for undirected graphs; we might also link the two occurrences of an edge (u, v), the first in u's adjacency list and the second in v's.

In connection with adjacency matrices we note the following well-known theorem. This concerns the kth matrical product, A^k, of the adjacency matrix, defined inductively as follows:

$$A^k(i,j) = \sum_{s=1}^{n} A^{k-1}(i,s)\, A(s,j)$$

where

$$A^1(i,j) = A(i,j)$$

Fig. 1.13. (*a*) A digraph G_1, its adjacency matrix A_1 and adjacency lists with their tabular representation T_1. (*b*) An undirected graph G_2, its adjacency matrix A_2 and adjacency lists with their tabular representation T_2.

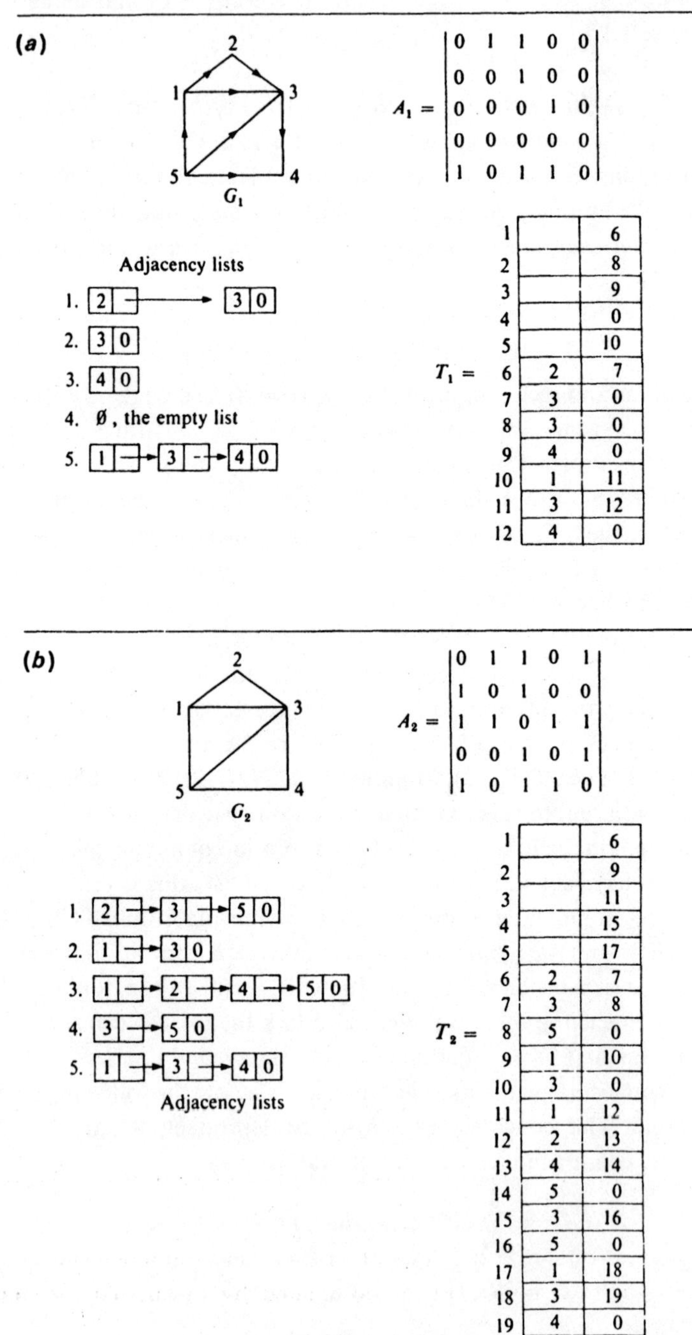

$$A_1 = \begin{vmatrix} 0 & 1 & 1 & 0 & 0 \\ 0 & 0 & 1 & 0 & 0 \\ 0 & 0 & 0 & 1 & 0 \\ 0 & 0 & 0 & 0 & 0 \\ 1 & 0 & 1 & 1 & 0 \end{vmatrix}$$

$T_1 =$

1		6
2		8
3		9
4		0
5		10
6	2	7
7	3	0
8	3	0
9	4	0
10	1	11
11	3	12
12	4	0

$$A_2 = \begin{vmatrix} 0 & 1 & 1 & 0 & 1 \\ 1 & 0 & 1 & 0 & 0 \\ 1 & 1 & 0 & 1 & 1 \\ 0 & 0 & 1 & 0 & 1 \\ 1 & 0 & 1 & 1 & 0 \end{vmatrix}$$

$T_2 =$

1		6
2		9
3		11
4		15
5		17
6	2	7
7	3	8
8	5	0
9	1	10
10	3	0
11	1	12
12	2	13
13	4	14
14	5	0
15	3	16
16	5	0
17	1	18
18	3	19
19	4	0

Theorem 1.4. $A^k(i,j)$ is the number of (non-simple) paths from i to j, containing k edges.

Proof. By induction on k. If $k = 1$ then $A^k(i,j) = 1$ if (i,j) exists and is zero otherwise. Thus we have a basis for the induction. We assume that the theorem is true for all powers of A less than the kth. Now, by definition:

$$A^k(i,j) = \sum_{s=1}^{n} A^{k-1}(i,s)\, A(s,j)$$

and by the induction hypothesis $A^{k-1}(i,s)$ is the number of paths from i to s and of length $(k-1)$. Thus $A^{k-1}(i,s)\, A(s,j)$ is the number of paths of length k from i to j which have (s,j) as a final edge. The sum is over all possible vertices adjacent to j and so the result follows. ■

Before coming to a description of depth-first searching we describe a matricial method to find the shortest paths between each pair of vertices in a weighted graph. This is in contrast to Dijkstra's algorithm we described earlier which finds the shortest paths from a specified vertex to all the others. The algorithm starts with a matrix W for which $W(i,j)$ is the weight, $w((v_i, v_j))$, of the edge (v_i, v_j). If v_i and v_j are distinct and $(v_i, v_j) \notin E$ then $w((v_i, v_j)) = \infty$, while $w((v, v)) = 0$. Then a series of matrices, $W_1, W_2, \ldots, W_n$ are constructed according to the following inductive rule:

$$W_k(i,j) = \min\,(W_{k-1}(i,j), (W_{k-1}(i,k) + W_{k-1}(k,j)))$$

where

$$W_0(i,j) = W(i,j)$$

W_n then provides the desired result according to the following theorem:

Theorem 1.5. $W_n(i,j)$ is the length of the shortest path from v_i to v_j.

Proof. We first show by induction on k, that $W_k(i,j)$ is the shortest path from v_i to v_j which passes only through vertices in the subset $\{v_1, v_2, \ldots, v_k\}$. If $k = 0$ then $W_k(i,j) = w((v_i, v_j))$, so that $W_k(i,j)$ is the length of the path (if it exists) from v_i to v_j which passes through no other vertex. We assume that the statement is true for $W_{k-1}(i,j)$. Now $W_k(i,j)$ is the smallest of $W_{k-1}(i,j)$ and $(W_{k-1}(i,k) + W_{k-1}(k,j))$. By the induction hypothesis $W_{k-1}(i,j)$ is the shortest path from v_i to v_j passing only through vertices in the subset $V' = \{v_1, v_2, \ldots, v_{k-1}\}$. If there is a shorter path which utilises v_k as well as the vertices of V' then its length, by the induction hypothesis must be $(W_{k-1}(i,k) + W_{k-1}(k,j)$. Thus the induction step follows.

When W_n has been constructed V' includes every vertex of the graph and so the theorem follows. ■

Figure 1.14 shows that this algorithm can be implemented to run in $O(n^3)$-time. The complexity is dominated by the nested for statements in lines 2 to 5. Notice that the space-complexity of the algorithm as outlined in figure 1.14 can be considerably improved (without detriment to the time-complexity) by recognising that W_k is only required for the computation of W_{k+1} and not for $W_{k+2}, \ldots, W_n$.

Fig. 1.14

```
1. Initialise W_0
2. for k = 1 to n do
3.         for i = 1 to n do
4.                 for j = 1 to n do
5.                         W_k(i,j) ← min ((W_{k-1}(i, k) + W_{k-1}(k, j)), W_{k-1}(i, j))
6. Output W_n
```

1.3.2. Depth-first searching

Most graph algorithms require a systematic method of visiting the vertices of a graph. A *depth-first search* (*DFS*) is just such a method which, as we shall see, has certain characteristics making some especially efficient algorithms possible.

For the time being we concern ourselves with undirected graphs only. Suppose then that in a depth-first search of an undirected graph we are currently visiting vertex v. The general step in the search then requires that we next visit a vertex adjacent to v which has not yet been visited. If no such vertex exists then the search returns to the vertex visited just before v and the general step is repeated until every vertex in that component of the graph has been visited. Such a search cannot revisit a vertex except by returning to that vertex via edges that have been used since the previous visit. Hence the edges traversed in a depth-first search form a spanning-tree for each separate component of the graph. This set of trees is called a depth first spanning forest, F. Thus a *DFS* partitions the edges E into two sets, F and $B = E-F$. The edges in B are called, for reasons which shall become evident, *back-edges*.

Before providing an example of a *DFS* of a graph we describe the method in terms of our algorithmic language. This is naturally achieved through the recursive procedure employed in figure 1.15. The input to this program consists of an adjacency list $A(v)$ for each vertex v of G. The output consists of the edge-set F. The algorithm uses a label $DFI(v)$ for each vertex v. Initially $DFI(v) = 0$, but on termination $DFI(v)$ is the order in which v was visited in the search. We shall call $DFI(v)$ the *depth-first index* of v. This ordering of the vertices is important for later algorithms and is best thought of as a renaming of the vertices. For a connected undirected graph line 11 of the algorithm could be omitted.

Fig. 1.15. A depth-first search of $G = \{A(v)|v \in V\}$ where $A(v)$ is the adjacency list for v.

```
 1. procedure DFS(v)
            begin
 2.         DFI(v) ← i
 3.         i ← i+1
 4.         for all v' ∈ A(v) do
 5.               if DFI(v') = 0 then
                    begin
 6.                 F ← F ∪ {(v v')}
 7.                 DFS(v')
                    end
            end of DFS
 8. i ← 1
 9. F ← ∅
10. for all v ∈ V do DFI(v) ← 0
11. while for some u, DFI(u) = 0 do
12.     DFS(u)
13. output F
```

The particularly efficient algorithms to be described in section 1.3.3 make use of the efficiency of *DFS*, which is established below, and the characteristics of the algorithm outlined in theorems 1.6 and 1.7.

Theorem 1.6. Following a depth-first search of an undirected graph each back-edge (u, v), connects an ancestor to a descendant.

Proof. We can, without loss of generality, presume that in a *DFS* u is visited before v. Thus $DFS(u)$ is called before $DFS(v)$ and $DFI(v) = 0$ when u is visited. All those vertices visited during the execution of $DFS(u)$ become descendants of u. Since u is in v's adjacency list, $DFS(u)$ will not terminate before v has been visited and so the theorem follows. ■

Figure 1.16 shows an application of the *DFS* algorithm and an illustration of theorem 1.6.

The complexity of the *DFS* algorithm is $O(\max(n, |E|))$ as follows. For each $v \in V$, $DFS(v)$ is called only once because after the first execution $DFI(v) = 0$. Apart from recursive calls of *DFS*, the time spent by $DFS(v)$ is proportional to $d(v)$, or for directed graphs $d^+(v)$. Thus calls of *DFS* take a total time proportional to $|E|$. On the other hand, line 10 requires $O(n)$ steps as does the search for successive components of the graph in line 11. Line 13 requires $O(|E|)$ steps. The result therefore follows.

Let us now suppose that a *directed* graph is subjected to the algorithm. In this case the edges of F form a spanning out-forest of the graph. Edges

Fig. 1.16. An example illustrating an application of the *DFS* algorithm. In lines 5 and 11 of figure 1.15 it is presumed that within the adjacency lists the vertices are ordered numerically according to their labels. (*a*) The two-component graph whose adjacency lists, when input to the *DFS* algorithm, produce the output below. (*b*) The spanning-forest *F* (in solid lines) output from the *DFS* algorithm. The back-edges are shown by dashed lines. (*c*) The depth-first order of visiting vertices.

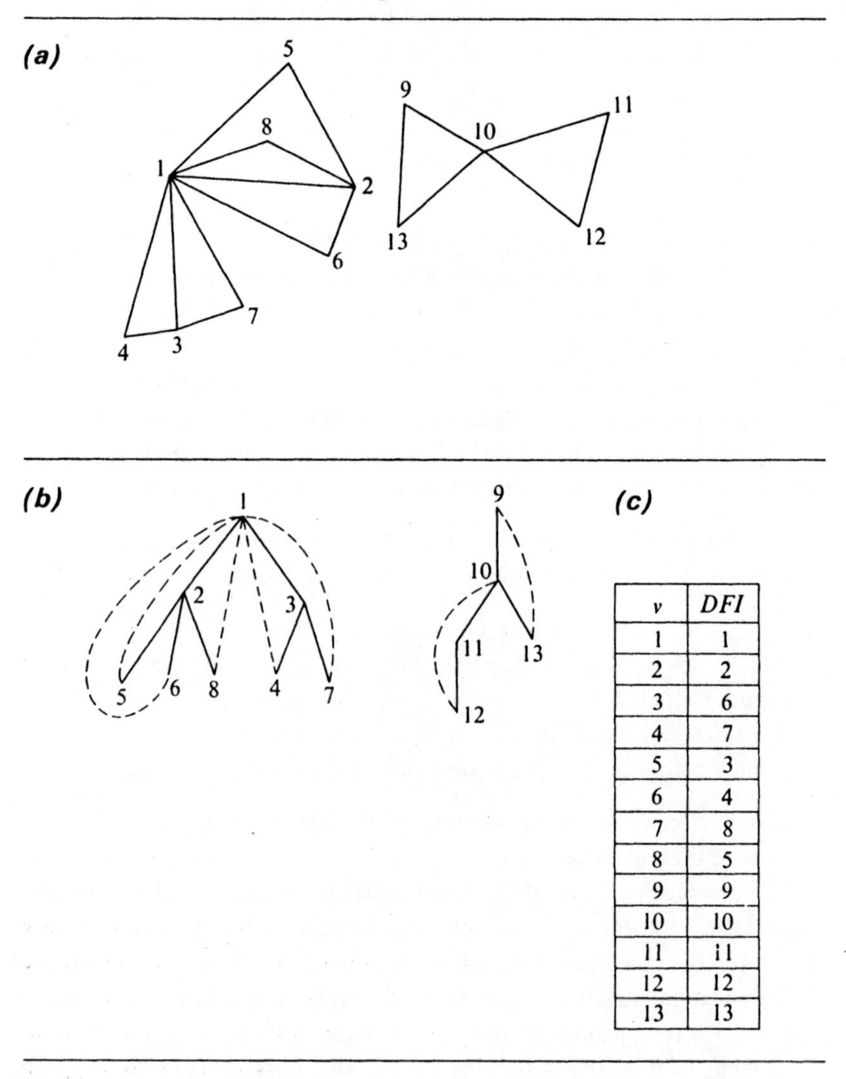

v	*DFI*
1	1
2	2
3	6
4	7
5	3
6	4
7	8
8	5
9	9
10	10
11	11
12	12
13	13

Fig. 1.17. An example illustrating an application of the *DFS* algorithm to a digraph. (*a*) A digraph which subjected to the *DFS* algorithm produces the output below. (*b*) The spanning out-forest F shown in solid lines. Back-, forward- and cross-edges are shown by dashed lines. (*c*) The depth-first order of visiting vertices.

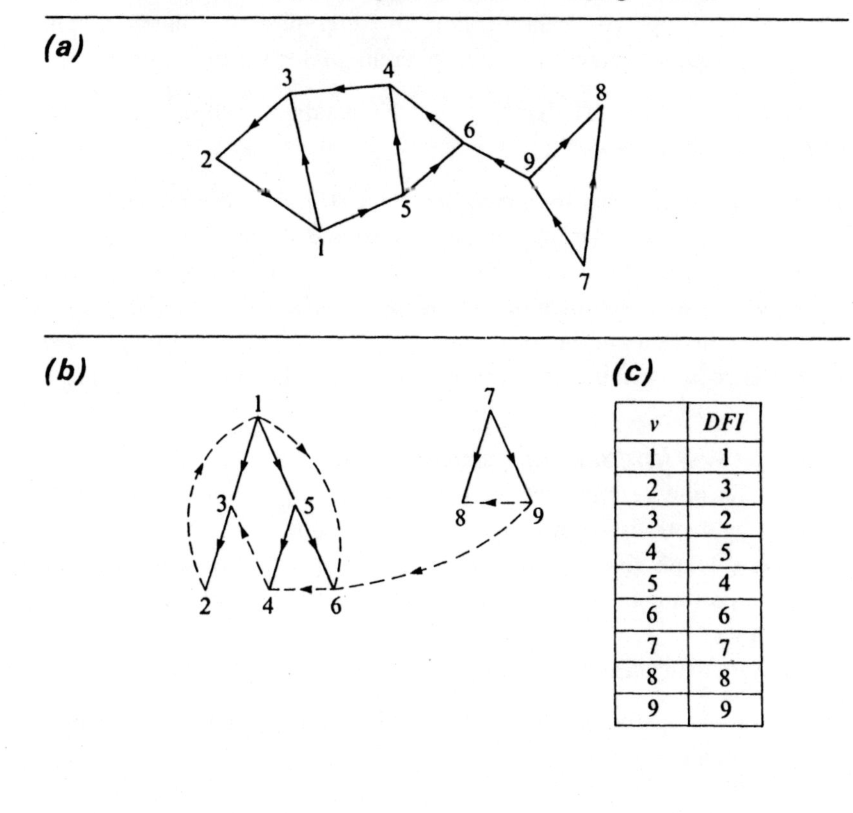

v	*DFI*
1	1
2	3
3	2
4	5
5	4
6	6
7	7
8	8
9	9

$F = \{(1, 3), (3, 2), (1, 5), (5, 4), (5, 6), (7, 8), (7, 9)\}$ $B_1 = \{(2, 1)\}$
$C = \{(4, 3), (6, 4), (9, 8), (9, 6)\}$ $B_2 = \{(1, 6)\}$

can only be added to F if they are directed away from the current vertex being visited. If no such edge exists to an unvisited vertex from those already visited, then the next vertex to be visited (if one exists) becomes the root of an out-tree. Figure 1.17 illustrates such an application of *DFS*. Notice that the search partitions the edges of the digraph into four types:

(i) a set of spanning-out forest edges, F,

(ii) a set of back-edges, B_1, which are directed from descendants to ancestors,

(iii) a set of forward-edges, B_2, which are directed from ancestors to descendants,
(iv) a set of cross-edges, C, which connect two vertices neither of which is a descendant of the other.

Despite this apparent complication for digraphs as compared with undirected graphs, there is a useful theorem analogous to theorem 1.6:

Theorem 1.7. Following a depth-first search of a digraph, if (u, v) is a cross-edge, then $DFI(u) > DFI(v)$.

Proof. Assume, contrary to the theorem, that $DFI(u) < DFI(v)$, that is, u is visited before v. Now v is in u's adjacency list and so must be visited within $DFS(u)$. Consider the possible type of the edge (u, v). If $DFI(v)$ is assigned when (u, v) is explored then (u, v) must be a tree edge. Otherwise v is first visited as a descendant, but not a son, of u. Then (u, v) must be a forward-edge. Hence (u, v) cannot be a cross-edge and so we have a contradiction. ■

1.3.3. Two linear-time algorithms

We now describe our first example of an algorithm made especially efficient by depth-first searching. This algorithm finds all the blocks of an undirected graph given its adjacency lists as input. Theorem 1.6 is crucial to this algorithm because the following observations can be made as a result of it:

If v is an articulation point then:

(*a*) If v is the root of a tree in the DFS spanning forest then v has more than one son.
(*b*) If v is not the root of a tree in the DFS spanning forest then v has a son v' such that no descendant of v' (which includes v') is connected by a back-edge to a proper ancestor of v.

These observations are illustrated in figure 1.16(*b*), where $v = 1$ is both a root and an articulation point and where $v = 10$ is not a root but is an articulation point.

In order to identify the blocks of a graph we need to identify its articulation points and the above observations can be used to do this. For the purpose of encoding (*b*) we associate a parameter $P(v)$ with each vertex v. If the vertices are labelled according to the order in which they are visited in a depth-first search, that is, by $DFI(v)$, then $P(v)$ is defined to be the smallest of v and those vertices which are connected by a back-edge to a descendant of v (including v). The maximum value of $P(v)$ is clearly v. Given this definition we can restate observation (*b*):

If v is an articulation point then:

(b') If v is not the root of a tree in the *DFS* spanning forest, then v has a son v' such that $P(v') \geqslant DFI(v)$.

We use the *DFS* algorithm as a basis for the block-finding algorithm and embed within it a calculation of $P(v)$. Since the *DFS* algorithm consists essentially of a recursive procedure, we need a recursive means to evaluate $P(v)$. This is provided by:

$$P(v) = \min(\{DFI(v)\} \cup \{P(v') | v' \text{ is a son of } v\} \cup \{DFI(v') | (v, v') \in B\})$$

Figure 1.18 incorporates this within the *DFS* algorithm. Line 3 of this depth-first search for blocks (*DFSB*) algorithm initialises $P(v)$ to its maximum possible value $DFI(v)$. Line 11 updates $P(v)$ if a son v' is found such that $P(v) \geqslant P(v')$ and again $P(v)$ is updated in line 12 if an appropriate back-edge is found. The articulation points are identified through line 10 whenever a vertex v is found such that $P(v') \geqslant DFI(v)$ for some son v'.

Fig. 1.18. The depth-first search for blocks algorithm.

```
 1.  procedure DFSB(v)
         begin
 2.      DFI(v) ← i
 3.      P(v) ← DFI(v)
 4.      i ← i+1
 5.      for all v' ∈ A(v) do
              begin
 6.           stack (v, v') if it has not already been stacked
 7.           if DFI(v') = 0 then
                   begin
 8.                father (v') ← v
 9.                DFSB(v')
10.                if P(v') ⩾ DFI(v) then pop and output the stack up
                                          to and including (v, v')
11.                P(v) ← min (P(v), P(v'))
                   end
12.           else if v' ≠ father (v) then P(v) ← min (P(v), DFI(v'))
              end
         end of DFSB
13.  i ← 1
14.  empty the stack
15.  for all v ∈ V do DFI(v) ← 0
16.  while for some v, DFI(v) = 0 do
17.           DFSB(v)
```

The *DFSB* algorithm incorporates a stack from which the set of edges of a block is popped as soon as it is found. Line 14 initialises the stack and line 6 stacks edges. We need to show that when a vertex v and a son v' are found for which $P(v') \geqslant DFI(v)$, then those edges on the stack, above and including (v, v'), define a block. This is easily shown by induction on the number of blocks B in a graph. If $B = 1$, then the only vertex v for which $P(v') \geqslant DFI(v)$ is the root of the *DFS* tree. When this is established every edge of the graph is on the stack with (v, v') at the bottom. We thus have a basis for our induction. As our induction hypothesis we assume that the statement is true for all graphs with less than B blocks. Now consider a graph with B blocks and let v be the *first* vertex for which it is found that $P(v') \geqslant DFI(v)$. No edges have been removed from the stack and those above (v, v') must be incident with vertices which are descendants of v'. Since v is an articulation point with no descendant which is an articulation point, those edges above and including (v, v') on the stack can only define the block containing (v, v'). When the edges of this block are removed from the stack, the algorithm behaves precisely as it would for the graph with $(B-1)$ blocks obtained by deleting the blocks containing (v, v'). This completes the inductive step of the proof.

Before providing an illustration of this algorithm we point out that if v is the root of a *DFS* tree then for every son v' of v we have $P(v') \geqslant DFI(v)$. This ensures that whenever v is revisited in a *DFS* search for blocks, the edges of the block containing (v, v') are removed from the stack. Thus the case when v is both a root *and* an articulation point is automatically accommodated.

Figure 1.19 shows an application of the depth-first search for blocks algorithm. In (*a*) the graph subjected to the algorithm is shown as are the spanning-tree, and the values of $DFI(v)$ and $P(v)$ found during the course of computation. In (*b*) we illustrate the state of the stack just before and just after the three occasions in which a vertex v is found for which $P(v') \geqslant DFI(v)$ for some son v'. We also indicate within which of the recursive calls of the *DFSB* procedure these occur.

The complexity of the depth-first search for blocks algorithm is $O(\max(n, |E|))$. This follows by a simple extension of the argument used for the complexity of the *DFS* algorithm. The only complication arises from the use of a stack. Clearly, however, the total time required over all calls of the *DFSB* procedure to stack and to subsequently pop edges is $O(|E|)$ and so the result follows. Thus the algorithm is especially efficient, operating within linear-time. This efficiency is achieved through the efficiency of the *DFS* algorithm and the particular characteristic expressed in theorem 1.6.

Fig. 1.19. Illustrating an application of the depth-first search for blocks algorithm of figure 1.18.

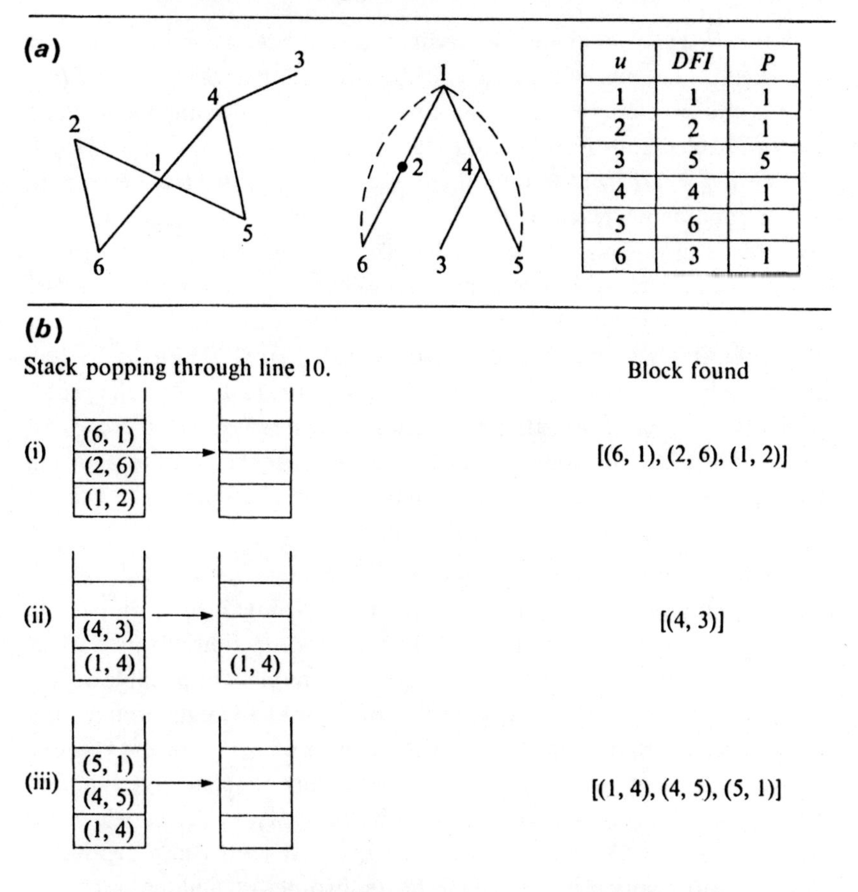

u	DFI	P
1	1	1
2	2	1
3	5	5
4	4	1
5	6	1
6	3	1

(i) Occurs within *DFSB*(1) after completion of *DFSB*(2), which itself contains a nested call of *DFSB*(6).

(ii) Occurs within *DFSB*(4) (which is nested within *DFSB*(1)) after completion of *DFSB*(3).

(iii) Occurs within *DFSB*(1) after completion of *DFSB*(4), which itself contains a call of *DFSB*(3) and a call of *DFSB*(5).

We now come to our second example of an algorithm made especially efficient by depth-first searching. This algorithm determines the strongly connected components of a digraph. The algorithm depends crucially upon theorem 1.8, which itself utilises theorem 1.7.

Theorem 1.8. If $G_i = (V_i, E_i)$ is a strongly connected component of the digraph G and if $F = (V_F, E_F)$ is a depth-first spanning forest, then $T_i = (V_i, E_i \cap E_F)$ is a tree.

Proof. We first show that any two vertices $u, v \in V_i$ have a common ancestor in V_i. Let us assume, without loss of generality, that $DFI(u) < DFI(v)$. Since u and v belong to the same strongly connected component, there exists a directed path P from u to v. We define w to be a vertex on P such that $DFI(w) < DFI(x)$ for any other vertex x on P. If we trace P from u, then as soon as we reach w, P can only pass through vertices which are descendants of w. This is because edges from descendants of w to vertices which are not descendants of w are either cross-edges (and note theorem 1.7) or back-edges and both must be to vertices with smaller depth-first indices. Thus w is an ancestor of v. Also since $DFI(w) \leqslant DFI(u) < DFI(v)$, u can only be a descendant of w. This completes the first part of the proof.

Let r be the root of the subtree containing every vertex in V_i. If $x \in V_i$ and if y is on the tree path from r to x, then we complete the proof by showing that $y \in V_i$. This is obviously the case since there is a path from r to y along the tree path, and there is a path from y to r via x. ■

The root of the tree T_i in the statement of theorem 1.8 is called the *root of the strongly connected component* G_i, and we denote it by r_i. In passing we note that what vertex within G_i is its root is a function of which edges are tree edges. A given digraph has, of course, a number of possible depth-first spanning forests. The *DFS* algorithm of Figure 1.15 might produce any one of these depending upon the initial (input) numbering of the vertices.

Theorem 1.8 suggests a natural way to determine the strongly connected components of a digraph G. We find the roots, $r_1, r_2, \ldots, r_k$, which we conveniently order so that if $i < j$, then r_i is last visited in a depth-first traversal of G before r_j is last visited. From theorem 1.8 and that r_j cannot be a descendant of r_i if $DFI(r_i) > DFI(r_j)$, we deduce that G_i is the subgraph induced by those vertices which are descendants of r_i but which are not also descendants of $r_1, r_2, \ldots, r_{i-1}$.

In the same way that we defined the parameter $P(v)$ to help in the computational discovery of articulation points in undirected graphs, we define a parameter $Q(v)$ to help in the computational identification of the roots of the strongly connected components of a digraph. $Q(v)$ is defined as follows:

$$Q(v) = \min(\{DFI(v)\} \cup \{DFI(v') | (x, v') \text{ is in } B_1 \text{ or } C, x \text{ is a descendant of } v \text{ and the root, } r, \text{ of the strongly connected component containing } v' \text{ is an ancestor of } v\})$$

The value of this definition lies in the following theorem.

Theorem 1.9. In a digraph G, v is the root of a strongly connected component if and only if $Q(v) = DFI(v)$.

Proof. Notice that by definition $Q(v) \leqslant DFI(v)$.

We first show that if v is the root of a strongly connected component then $Q(v) = DFI(v)$. Suppose that, on the contrary, $Q(v) < DFI(v)$. Therefore there exists a vertex v', as in the definition of $Q(v)$, such that $DFI(v') < DFI(v)$. Now $DFI(r) < DFI(v')$ so that we have $DFI(r) < DFI(v)$. But r and v must belong to the same strongly connected component because there is a path from r to v and a path from v to r via (x, v'). Thus, since $DFI(r) < DFI(v)$, v cannot be the root of a strongly connected component. This is a contradiction and so we conclude that $Q(v) = DFI(v)$.

We now only need show that if v is not the root of a strongly connected component then $Q(v) < DFI(v)$. Let us assume, however, that $Q(v) = DFI(v)$, so that no vertex v', as described in the definition of $Q(v)$, should exist for which $DFI(v') < DFI(v)$. Since v is not the root, some other vertex r must be the root. Then there must exist a path P from v to r which contains a first vertex (maybe r) which is not a descendant of v. Let this vertex be v'. Clearly, r and v' belong to the same strongly connected component. The edge of P incident to v' is in B_1 or C. Thus $DFI(v') < DFI(v)$ which is a contradiction. Hence $Q(v) < DFI(v)$. ∎

Again we use the *DFS* algorithm as a basis and embed within it a calculation of $Q(v)$. As with $P(v)$ of the previous example, we require a recursive method of evaluation for $Q(v)$. This is provided by:

$$Q(v) = \min(\{DFI(v)\} \cup \{Q(v') \mid v' \text{ is a son of } v\} \cup \{DFI(v') \mid (v, v') \text{ is in } B_1 \text{ or } C \text{ such that the root of the strongly connected component containing } v' \text{ is an ancestor of } v\})$$

Figure 1.20 incorporates this within the *DFS* algorithm. The modified procedure $DFSSCC(v)$, depth-first search for strongly connected components, includes a stack upon which vertices are placed in line 5. An array called *stacked* is used to record which vertices are on the stack. Line 3 initialises $Q(v)$ to its maximum possible value and line 9 updates $Q(v)$ if a son of v, v', is found such that $Q(v') < Q(v)$. Line 10 further updates $Q(v)$ if an edge (v, v') in B_1 or C is found such that the root of the strongly connected component containing v' is an ancestor of v. Notice that at line 10 $DFI(v') \neq 0$ and so v' has been previously visited and since $DFI(v') < DFI(v)$ for the update to take place, (v, v') cannot be a forward-edge. Also, since v' is stacked, the root of the strongly connected component containing v' has yet to be identified and so, because of the order in which roots are identified, must be an ancestor of v. Line 11 identifies roots and, again because of theorem 1.8 and the order of identifying roots, those vertices above and including the root on the stack must induce a strongly connected component. These are then removed from the stack and output.

Fig. 1.20. The depth-first search for strongly connected components algorithm.

```
 1. procedure DFSSCC(v)
        begin
 2.     DFI(v) ← i
 3.     Q(v) ← DFI(v)
 4.     i ← i+1
 5.     put v on the stack and set stacked (v) ← true
 6.     for all v' ∈ A(v) do
 7.         if DFI(v') = 0 then
                begin
 8.             DFSSCC(v')
 9.             Q(v) ← min (Q(v), Q(v'))
                end
10.         else if DFI(v') < DFI(v) and stacked (v')
                   then Q(v) ← min (Q(v), DFI(v'))
11.     if Q(v) = DFI(v) then pop and output the stack up to
            and including v, for each popped vertex u reset
            stacked (u) ← false
        end of DFSSCC
12. i ← 1
13. empty the stack
14. for all v ∈ V do begin DFI(v) ← 0, stacked (v) ← false end
15. while for some u, DFI(u) = 0 do
16.         DFSSCC(u)
```

Figure 1.21 shows an application of the depth-first search for strongly connected components algorithm. In (*a*) the graph subjected to the algorithm is shown as are the spanning forest and the values of $DFI(v)$ and $Q(v)$ found during the course of computation. In (*b*) we illustrate the state of the stack just before a vertex is found for which $Q(v) = DFI(v)$ and just after the vertices of a strongly connected component have been popped from it. We also indicate within which of the recursive calls of *DFSSCC* the strongly connected components are found.

The *DFSSCC* algorithm operates within linear time, having a complexity $O(\max(n, |E|))$. This follows by a similar argument to that employed for the *DFSB* algorithm. Notice that the use of the array called *stacked* enables line 11 of figure 1.20 to check whether v' is on the stack or not in one step. This avoids unnecessary enhancement of the complexity through a search of the stack in line 11.

Fig. 1.21. Illustrating an application of the depth-first search for strongly connected components algorithm of figure 1.20.

(a)

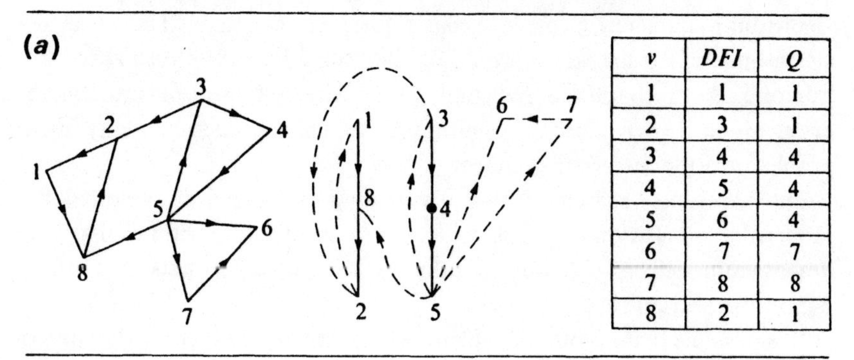

v	*DFI*	*Q*
1	1	1
2	3	1
3	4	4
4	5	4
5	6	4
6	7	7
7	8	8
8	2	1

(b)

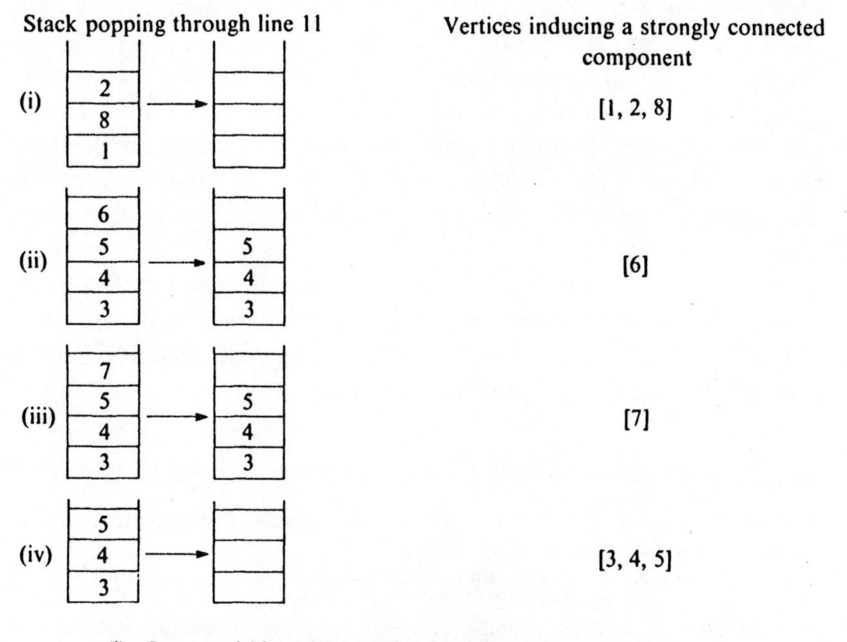

(i) Occurs within *DFSSCC*(1) after completion of *DFSSCC*(8) which contains a nested call of *DFSSCC*(2).

(ii) Occurs within *DFSSCC*(6) which is nested successively within *DFSSCC*(5), *DFSSCC*(4) and *DFSSCC*(3).

(iii) Occurs within *DFSSCC*(7) which is called immediately after *DFSSCC*(6).

(iv) Occurs within *DFSSCC*(3) which is called after the completion of *DFSSCC*(1).

1.4 Summary and references

The basic definitions of this chapter concerning graphs and algorithmic efficiency provide a basis for later chapters. The ideas of algorithmic efficiency which we briefly described were first formalised by Edmonds[1]. In chapter 8 we shall pursue the question of intractable problems in a formalised way, although we shall encounter many such problems in the intervening chapters.

The depth-first search of a depth of a graph was first exploited by Hopcroft and Tarjan in [2] and [3], and it is their algorithms that are described in section 1.3.3. Several other linear-time algorithms also utilise depth-first searching.

It is possible that no other problem in graph theory has received as much attention as one used as an example in this chapter. This is the problem of finding shortest paths. The problem can be posed with different constraints and for each case there can be an appropriate algorithm. The review by Dreyfus[4] is recommended. Dijkstra's algorithm,[5] described in the test, applies to directed or to undirected graphs with non-negative edge-weights. The $O(n^3)$ algorithm for all pairs of vertices described in the text is due to Floyd[6] and is based on work by Warshall.[7] For this problem note Spira.[13] Exercises 1.8 and 1.15 are also about shortest path algorithms.

As far as general reading is concerned, Aho, Hopcroft & Ullman,[8] Deo[9] and Even[10] provide particularly useful elaboration for this chapter.

To some extent the problems that follow extend material in this chapter. Exercises 1.14 to 1.16 are particularly recommended.

[1] Edmonds, J. 'Paths, trees and flowers', *Canad. J. of Maths*, **17**, 449–67 (1965).
[2] Hopcroft, J. & Tarjan, R. 'Algorithm 447: efficient algorithms for graph manipulation', *CACM*, **16**, 372–78 (1973).
[3] Tarjan, R. 'Depth-first search and linear graph algorithms', *SIAM. J. Comput*, **1**, 146–60 (1972).
[4] Dreyfus, S. E. 'An appraisal of some shortest path algorithms', *J. Operations Research*, **17** (3), 395–412 (1969).
[5] Dijkstra, E. W. 'A note on two problems in connection with graphs', *Numerische Math.*, **1**, 269–71 (1959).
[6] Floyd, R. W. 'Algorithm 97: Shortest path', *CACM*, **5**, 345 (1962).
[7] Warshall, S. 'A theorem on Boolean matrices', *JACM*, **9**, 11–12 (1962).
[8] Aho, A. V., Hopcroft, J. E. & Ullman, J. D. *The Design and Analysis of Computer Algorithms*. Addison-Wesley (1974).
[9] Deo, N. *Graph Theory with Applications to Engineering and Computer Science*. Prentice-Hall (1974).
[10] Even, S. *Graph Algorithms*. Computer Science Press (1979).
[11] Moore, E. F. 'The shortest path through a maze', *Proc. Internat. Symp. Switching Th.*, 1957, Part II, Harvard University Press, pp. 285–92 (1959).

[12] Munro, J. I. 'Efficient determination of the transitive closure of a directed graph', *Information Processing Letters*, **1**, 56–8 (1971).
[13] Spira, P. M. 'A new algorithm for finding all shortest paths in a graph of positive arcs in average time $O(n^2 \log ^2 n)$', *SIAM J. Computing*, **2**, 28–32 (1973).

EXERCISES

1.1. Draw every simple graph with n vertices, $1 \leqslant n \leqslant 4$.

1.2. Show that any two of the following regular bipartite graphs are isomorphic.

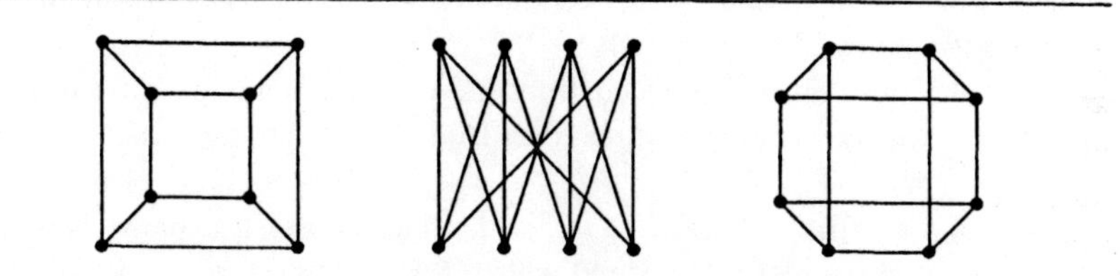

1.3. Show that in a disconnected graph there must be a path from any vertex of odd-degree to some other vertex of odd-degree.
(Use theorem 1.1.)

1.4. Show that any connected graph G with n vertices and $(n-1)$ edges must be a tree.
(Show that the assumption that G is not a tree, that is it contains a circuit, leads to a contradiction that G cannot be connected.)

1.5. Show that in a binary tree with n vertices:
(*a*) n is always odd.
(*b*) The number of vertices of degree 1 is $\frac{1}{2}(n+1)$.

1.6. Show that in a connected simple graph with n vertices, the number of edges $|E|$ satisfies:

$$(n-1) \leqslant |E| \leqslant \tfrac{1}{2}n(n-1)$$

(The lower limit corresponds to a tree, and the upper limit to a complete graph.)

1.7. Show that if a simple graph has more than $\frac{1}{2}(n-1)(n-2)$ edges, then it must be connected.
(First show, by induction on k, that a simple graph with k components has at most $\frac{1}{2}(n-k)(n-k+1)$ edges.)

1.8. Figure 1.14 illustrates an $O(n^3)$-algorithm to find the shortest distance between every pair of vertices in a weighted graph. Describe an alternative algorithm for this task which procedurally incorporates Dijkstra's algorithm. What is the complexity of your algorithm?

1.9. A binary relation R is an *equivalence* relation on the set S if R is:
(*a*) *reflexive*, that is, aRa for all $a \in S$,
(*b*) *symmetric*, that is, aRb implies bRa for all $a, b \in S$,

and

(*c*) *transitive*, that is, aRb and bRc implies that aRc.

It is easy to see that an equivalence relation on S partitions S into disjoint subsets called *equivalence classes*. Two elements of S are in the same equivalence class if and only if they are related by R.

Show that the relations of:

(*a*) 'connected to' – in an undirected graph,

and

(*b*) 'strongly connected to' – in a directed graph

are examples of equivalence relations on the vertex-set of a graph.

1.10. For the undirected weighted graph shown and as in the matricial algorithms of section 1.3.1:

(*a*) Construct A^3 and so confirm that there are ten non-simple distinct paths consisting of three edges from v_3 to v_5. Describe each of these paths.

(*b*) Construct $W_0, W_1, \ldots, W_6$, so finding the shortest paths between each pair of vertices and which, for W_i, only have $v_1, v_2, \ldots, v_i$ as internal vertices.

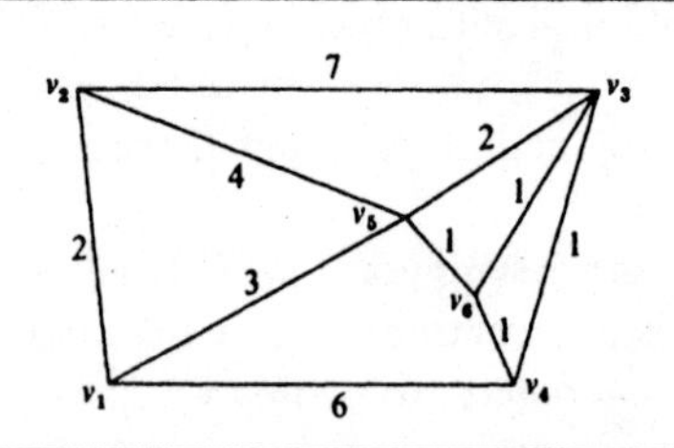

1.11. The *transitive closure* of a digraph $G = (V, E_1)$ is a digraph $T_G = (V, E_2)$ such that if there is a path from u to v in G, then $(u, v) \in E_2$. Clearly E_1 is a subset of E_2. Describe an algorithm to construct the adjacency matrix $A(T_G)$ of T_G from the adjacency matrix of G. For obvious reasons $A(T_G)$ is sometimes called the *reachability matrix* of G.

(An $O(n^3)$ algorithm can easily be obtained by modifying the algorithm for finding the shortest paths between all pairs of vertices which is described in the text. Note that Munro has described faster algorithms.[12])

1.12. Given the reachability matrix of a digraph (see the previous question design an $O(n^2)$-algorithm to identify its strongly connected components. (If $A_{ij}(T_G) = A_{ji}(T_G) = 1$ then v_i and v_j belong to the same strongly connected component.)

1.13. The *condensation* C_G of a digraph G is a digraph obtained from G by replacing each of its strongly connected components by a vertex and each non-empty set of edges from one strongly connected component to another by a single edge. For example:

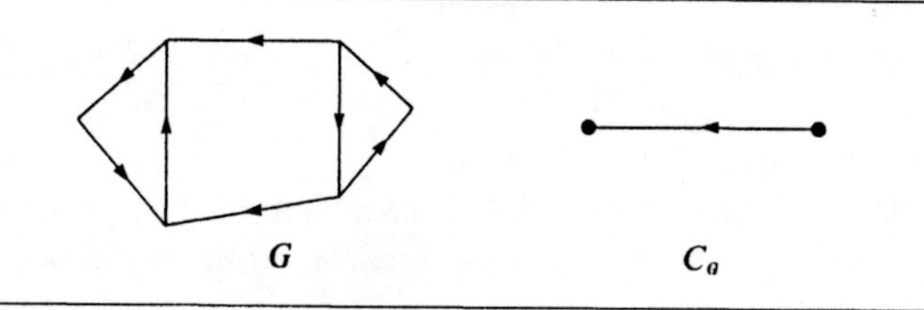

Show that the condensation of a directed graph cannot contain a (directed) circuit.

1.14. Given the adjacency lists $A(v)$ for each vertex v of a connected graph (directed or undirected), the following algorithm conducts a *breadth-first search*. On completion of the search each vertex has acquired a *breadth-first index* (*BFI*) indicating the order in which the vertex was visited. Vertex u is visited first and $BFI(u) = 0$. Hand-turn the algorithm on a small graph of your choice. Notice that use is made of a *queue* which is a data structure in which items are removed in the same order that they are added. Such a structure is also known as *FIFO* (first-in, first-out) store. Show that any graph will be traversed by the algorithm in $O(\max(n, |E|)$ steps. Why is such a traversal called a breadth-first search?

```
 1. for all v ∈ V do BFI(v) ← 0
 2. i ← 1, BFI(u) ← 1
 3. add u to the queue
 4. while the queue is not empty do
        begin
 5.     remove a vertex from the queue, call it w
 6.     for all v ∈ A(w) do
 7.         if BFI(v) = 0 then
                begin
 8.             BFI(v) ← i+1
 9.             i ← i+1
10.             add v to the queue
                end
        end
```

The depth-first search algorithm of figure 1.15 constructs a depth-first spanning-tree for a connected graph. Modify the above algorithm to construct a *breadth-first spanning tree.*

1.15. Just as depth-first searching is a suitable way to traverse a graph for some algorithms, so is breadth-first searching as described in the previous question. For example, an algorithm which finds the lengths of the shortest paths from a vertex u to every other vertex of an unweighted connected graph can be obtained by editing the algorithm of the previous question as follows:

(*a*) Between lines 5 and 6 insert:

if $BFI(w) \neq i$ **then** $i \leftarrow i+1$.

(*b*) Delete line 9,

(*c*) Throughout replace *BFI* by *L*.

Show that on completion of this breadth-first search for shortest paths (*BFSSP*) algorithm, due to Moore,[11] $(L(v)-1)$ is the shortest path from u to v.

(Use induction on $L(v)$.)

Show that the *BFSSP* algorithm has complexity $O(\max(n, |E|)$ and therefore for sparse graphs (in which $|E| \ll n^2$) it is more efficient than Dijkstra's algorithm. However, notice that the *BFSSP* algorithm cannot be used for weighted graphs. Also note the implementation of Dijkstra's algorithm outlined in exercise 1.16(*b*) which has an improved complexity for sparse graphs.

1.16. Here we introduce a data structure known as a *priority queue* and illustrate one use of it. Another illustration can be found in exercise 2.14.

(*a*) A *priority queue* is an abstract data type in which a priority is associated with each of its k elements. We can add an element to the data structure and we can delete (or remove) the element of lowest priority. Such a structure can be implemented in several ways. For example a sorted *or* an unsorted list will do. For the former the insertion operation, and for the latter the priority deletion operation, will require $O(k)$-time. We can improve on this by using, for example, a *partially ordered* tree. As we shall see, $O(\log k)$-time is then sufficient for either operation.

We define a partially ordered tree to be a binary tree with the (priorities of the) data elements located at the vertices. The elements are arranged in partial order, by which we mean that the priority of any vertex is no greater than the priority of its sons. Moreover, the tree is as *balanced* as possible (path lengths from the root to the leaves differ by at most one) with leaves furthest from the root being arranged to the left. Such a tree is shown opposite (figure (*a*)).

Consider first the operation of removing the item of lowest priority. This item will be located at the root of the tree so that its removal no longer leaves us with a tree. To overcome this, the root is initially replaced with the rightmost element from the lowest level of the tree. In order to re-establish the partial ordering, this element is repeatedly exchanged with one of its sons (the one of least priority) until no son has lower priority than this element. Figure (*b*) shows this process for the tree of figure (*a*).

Now consider adding an element to the tree. We can do this by creating a new leaf at the lowest level and as far to the left as possible. Placing the new element may require the partial ordering to be re-

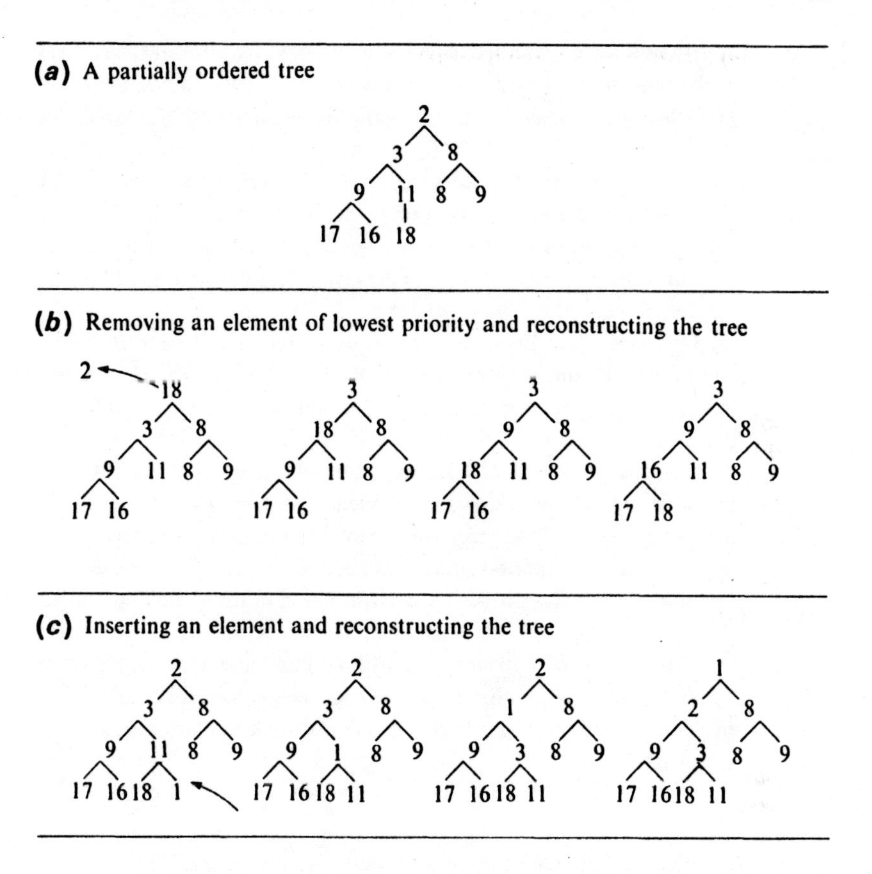

established. This is done by repeatedly exchanging the new element with its father so long as its father has higher priority. This is illustrated in figure (*c*) where an item with priority 1 is added to the tree of figure (*a*). In this example the new item eventually filters to the root of the tree.

Justify the following claims:

(i) In a partially ordered tree, the element of lowest priority is at the root of the tree.

(ii) A partially ordered tree is in fact re-established after the operations of adding an item and removing the one of least priority as we have described.

(iii) The complexity of adding an item and the complexity of removing one of lowest priority are both $O(\log_2 k)$.
(The complexity will be determined by the number of exchanges of elements required to re-establish the partial ordering. This is obviously bounded by the maximum path length from root to leaf.)

(iv) A priority queue of k items can be constructed in $O(k \log k)$-time if it is implemented using a partially ordered tree.

It is interesting to note that a partially ordered tree, as defined, can be very usefully represented in terms of an array. We number the vertices of the tree from top to bottom and within each level from left to right. Then using an array H, the ith vertex is located in $H[i]$. With this arrangement, the left son of $H[i]$ is in $H[2i]$ and the right son is in $H[2i+1]$. Moreover, the father of $H[i]$, if it exists, is in $H[i \textbf{ div } 2]$. Here $i \textbf{ div } 2$ is the integral part of $\frac{1}{2}i$. Such an array is called a *HEAP*.

(v) Making use of a *HEAP*, write detailed O(log k)-algorithms both for removing an element of lowest priority and for adding an an element to a priority queue.

(*b*) Consider again Dijkstra's algorithm of figure 1.10. Set up before line 4 a priority queue for the vertices in $(V-T)$. Line 5 can then simply be the priority deletion operation on this structure. Thus for all iterations of line 5, O(n log n)-time is required. For line 7, the only possible changes to the $L(v)$ are for those v adjacent to v'. These can be attended to by skipping down an adjacency list of v'. For *all* v' the total number of skips will be proportional to $|E|$. For each skip, apart from updating $L(v)$, the partial order will need to be restored as a result of this updating. This can be achieved in O(log n)-time, moving $L(v)$ up the tree as required. Thus overall, construct an $O((|E|+n) \log n)$ implementation of Dijkstra's algorithm. For large sparse graphs (in which $|E| \ll n^2/\log n$) this complexity represents a considerable saving in running time compared with the $O(n^2)$ implementation implied in the chapter.

2

Spanning-trees, branchings and connectivity

Trees are the most commonly occurring type of graph in models and computations of all kinds. Computer scientists are particularly familiar with these structures through the consideration of parse trees, tree searches, game trees and so on. In chapter 1 we defined trees and provided further characterisation of them through theorem 1.2.

Given an arbitrary graph, our interest in this chapter is with certain of its subgraphs which are trees. In the first half of the chapter we consider weighted graphs and digraphs. For these algorithms are described which find spanning-trees and forests of out-trees of optimal weight. In the second half of the chapter we show how spanning-trees play an important rôle in connection with the circuit space and with the separability of a graph. This leads naturally to a generalisation of the definitions of cut-edge and articulation point which we provided in chapter 1.

2.1 Spanning-trees and branchings

A spanning-tree of a connected undirected graph G is a subgraph which is both a tree and which contains all the vertices of G. As we saw in chapter 1, such a spanning-tree can be found in linear-time using, for example, a depth-first search, such as we described in section 1.3.2 or a breadth-first search as indicated in exercise 1.14. In this section we first describe an algorithm which solves a more general task. This is to find, given a weighted, connected and undirected graph, a spanning-tree of minimum weight. This problem may appear in a number of guises, the most common of which concerns the construction of a communication network, perhaps a road or a railway system linking a set of towns. Given the cost of directly connecting any two towns, the problem is to find a network at minimum cost and which provides some route between every two towns. The solution is the minimum-weight spanning-tree of the associated complete weighted graph. Because of this description the problem is often called the *connector problem.* The similar problem of finding a maximum weight spanning-tree can be solved by making a minor modification to the minimum weight spanning-tree algorithm we shall describe.

We shall also describe an algorithm to solve a similar but more difficult problem for digraphs. Given a weighted digraph the problem is to find a maximum- (or minimum-) weight forest of out-trees which is a subgraph of the digraph. Such a forest is called a maximum (or minimum) branching.

We complete this section with a description of how to count the spanning-trees of a graph.

2.1.1. Optimum weight spanning-trees

There are a number of algorithms known to solve the connector problem for undirected graphs. The best-known of these are due to Prim[1] which we describe here, and to Kruskal[2] which is outlined in exercise 2.4.

Prim's algorithm is described in figure 2.1. At each iterative stage of the algorithm a new edge e is added to T. Now, T is a connected subgraph of the minimum-weight spanning-tree under construction, and it spans a subset of vertices $V' \subset V$. The edge e is the edge of least weight connecting a vertex in $(V-V')$ to a vertex in V'. Initially V' contains some arbitrary vertex u. At each stage, the label $L(v)$, for each vertex v, records the edge of least weight from $v \in (V-V')$ to a vertex in V'. Thus each $L(v)$ is initialised to the weight $w((u, v))$ of the edge (u, v), provided $(u, v) \in E$. Otherwise $w((u, v)) = \infty$ if u and v are distinct, whilst $w((v, v)) = 0$. Line 8 of the algorithm updates the $L(v)$ whenever a new vertex w has been added to V'. The algorithm stops when $V' = V$ at line 4. An example of an application of Prim's algorithm is shown in figure 2.2.

The following theorem proves that Prim's algorithm works.

Theorem 2.1. Prim's algorithm finds a minimum-weight spanning-tree of a connected undirected graph G.

Proof. We prove by induction on the size of V' that T is a subtree of a minimum-weight spanning-tree of G spanning the vertices of V'.

As the basis for our induction, we note that the statement is trivially true when T and V' are initialised in lines 1 and 2 of the algorithm. We assume then, that it is true, whatever the value of $|V'|$, just before e is added to T in line 6. Now consider $(T+e)$. We first show that $(T+e)$ is a tree.

The edge e serves to connect w to a vertex of T. Since by the induction hypothesis T is connected, $(T+e)$ must be also. Now $(T+e)$ cannot contain a circuit because T does not and because one end-point of e, namely w, is of degree one in $(T+e)$. Thus $(T+e)$ is both connected and acyclic and must therefore be a tree.

Fig. 2.1. Prim's minimum-weight spanning-tree algorithm.

1. $T \leftarrow \varnothing$
2. $V' \leftarrow \{u\}$
3. **for all** $v \in (V-V')$ **do** $L(v) \leftarrow w((u, v))$
4. **while** $V' \neq V$ **do**
   begin
5. find a w for which $L(w) = \min\{L(v) | v \in (V-V')\}$ and denote the associated edge from V' to w by e
6. $T \leftarrow T \cup \{e\}$
7. $V' \leftarrow V' \cup \{w\}$
8. **for all** $v \in (V-V')$ **do**
   $L(v) \leftarrow$ **if** $w((v, w)) < L(v)$ **then** $w((v, w))$
   end

Fig. 2.2. An application of Prim's algorithm. For each iteration of the while-loop T becomes $T \cup \{e\}$ and V' becomes $V' \cup \{w\}$. Finally, T is the minimum-weight spanning-tree consisting of the heavily scored edges.

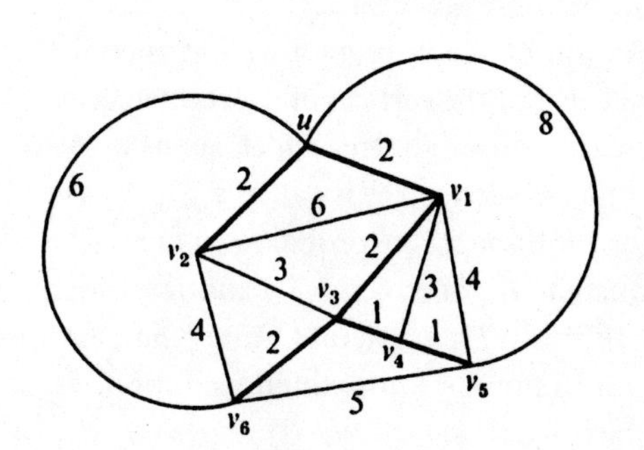

Iteration	w	e
1	v_1	(u, v_1)
2	v_2	(u, v_2)
3	v_3	(v_1, v_3)
4	v_4	(v_3, v_4)
5	v_5	(v_4, v_5)
6	v_6	(v_3, v_6)

We now show that $(T+e)$ is a subtree of a minimum-weight spanning-tree of G. By the induction hypothesis $T \subset T_M$, where T_M is some minimum-weight spanning-tree of G. Suppose that e is not an edge of T_M. Then by theorem 1.2 (T_M+e) contains a circuit C. One edge of C, namely e, connects a vertex in V' to a vertex in $(V-V')$. There is therefore another edge e' from V' to $(V-V')$ on C. If we now construct the tree $T'_M = (T_M+e)-e'$, we notice that $(T+e) \subset T'_M$. Moreover T'_M is a minimum-weight spanning-tree because:

$$w(T'_M) = w(T_M) + w(e) - w(e')$$

and from line 5 of the algorithm:

$$w(e) \leqslant w(e')$$

so that:

$$w(T'_M) = w(T_M)$$

Thus when e is added to T, $(T+e)$ is still a subtree of some minimum-weight spanning-tree.

On completion of the algorithm through line 4, $V' = V$ so that T spans G. ■

Prim's algorithm is an efficient one, as can easily be seen as follows. The while body, lines 5 to 8 of figure 2.1, is executed $(n-1)$ times. Within each execution both the computation of min $\{L(v) | v \in (V-V')\}$ at line 5 and the for statement of line 8 can be executed within $O(n)$ steps. Thus overall we have an $O(n^2)$-algorithm. More efficient algorithms are known. See, for example, Cheriton & Tarjan.[3] Note also exercise 2.14.

Prim's algorithm, as we have described it, finds a minimum-weight spanning-tree. It is easy to see that the simple modification of replacing min $\{L(v)\}$ by max $\{L(v)\}$ in line 5 will cause the algorithm to find a maximum-weight spanning-tree. The proof of this is obtained by replacing 'minimum' by 'maximum' throughout theorem 2.1.

The following is a generalisation of the minimum-weight spanning-tree problem. Given a proper subset, V', of the vertices of a weighted connected and undirected graph, find a minimum-weight tree which spans the vertices of V' and, if necessary, some others. Such a tree is called a *Steiner tree*. No efficient algorithm is known for the Steiner tree problem. In fact the related decision problem (Given a constant W, does G have a subtree which both spans V' and is of weight $\leqslant W$?) is *NP*-complete. Chang[4] has, however, described an efficient algorithm to find an approximate solution.

2.1.1. Optimum branchings

In the preceding subsection we were concerned to find optimal weighted trees for undirected graphs. Here we look at a similar problem for digraphs. We describe an algorithm due to Edmonds.[5] This finds a subgraph of a digraph which is a maximum-weight, which does not necessarily mean spanning, forest of out-trees called a *maximum branching*. We shall see that the same algorithm, with minor changes, can be used, to find a *minimum branching*.

The algorithm traverses the digraph examining vertices and edges. It places vertices in a so-called *vertex-bucket BV* as they have been examined, and edges in an *edge-bucket BE* if they have been provisionally chosen for the branching. Throughout the course of the algorithm *BE* always contains a branching, that is an acyclic collection of directed edges with at most one edge incident to any given vertex. The examination of a vertex v consists simply of choosing an edge of maximum *positive* weight e that is incident to v. Notice that no edge of negative weight would be chosen for a maximum branching (a digraph consisting of negative weighted edges only has a maximum branching of zero weight and no edges). The edge e is checked to see if it forms a circuit with the edges already in *BE*. If it does not then e is added to *BE* and a new vertex is examined. If is does then the graph is restructured by shrinking this circuit to a single vertex and assigning new weights to those edges which are incident to this new 'artificial' vertex. The process of examining vertices then continues until *BV* contains all the vertices of a final graph. It contains just these vertices, several of which may in general be 'artificial', because whenever a circuit is shrunk to form a new graph the edges and vertices of the circuit are removed from *BE* and *BV*. *BE* at this stage contains the edges of a maximal branching for the final graph. The reverse process of replacing in turn each of the artificially created vertices by its associated circuit then begins. At each replacement the choice of edges placed in *BE* is such that for the currently reconstructed graph *BE* contains the edges of a maximum branching. As we shall see, the crucial element of the algorithm is the rule for reassigning weights to edges when circuits are shrunk. It is this which forces the choice of edges to be included in the branching when the reconstruction phase is underway.

An outline of the algorithm is shown in figure 2.3. The original digraph, input to the algorithm, is $G_0 = (V_0, E_0)$ and $G_i = (V_i, E_i)$ is the graph obtained after the ith circuit C_i has been replaced by a single vertex u_i. Lines 1–13 inclusive generate the succession of graphs $G_1, G_2, \ldots, G_k$ provided one or more circuits have to be shrunk to artificial vertices. This process ceases when *BV* contains the edges of the current graph at line 4. We need to fill in the details by which G_i is constructed from G_{i-1} and by which *BE*, *BV* and some edge-weights are modified in lines 10 and 11.

As might be imagined G_i contains every vertex of G_{i-1} except for those in C_i. V_i also includes the new vertex u_i. E_i contains every edge of E_{i-1} except those with one or more end-points in C_i. We also add to E_i new edges as follows. For every edge $(x, y) \in E_{i-1}$ for which $x \notin C_i$ and $y \in C_i$ (or for which $x \in C_i$ and $y \notin C_i$), E_i contains an edge (x, u_i) (or an edge (u_i, y)). Any edge in G_i has the same weight as its corresponding edge in G_{i-1}

Fig. 2.3. Edmond's algorithm to find a maximum branching.

---

1. $BV \leftarrow BE \leftarrow \varnothing$
2. $i \leftarrow 0$
3. **if** $BV = V_i$ **then go to** 14
4. **for** some vertex $v \notin BV$ **and** $v \in V_i$ **do**
   begin
5. $BV \leftarrow BV \cup \{v\}$
6. find an edge $e = (x, v)$ such that
   $w(e) = \max\{w(y, v) | (y, v) \in E_i\}$
7. **if** $w(e) \leqslant 0$ **then go to** 3
   end
8. **if** $BE \cup \{e\}$ contains a circuit **then**
   begin
9. $i \leftarrow i+1$
10. construct G_i by shrinking C_i to u_i
11. modify BE, BV and some edge-weights
    end
12. $BE \leftarrow BE \cup \{e\}$
13. **go to** 3
14. **while** $i \neq 0$ **do**
    begin
15. reconstruct G_{i-1} and rename some edges in BE
16. **if** u_i was a root of an out-tree in BE **then**
17. $BE \leftarrow BE \cup \{e | e \in C_i \text{ and } e \neq e_0^i\}$
18. **else** $BE \leftarrow BE \cup \{e | e \in C_i \text{ and } e \neq \tilde{e}_i\}$
19. $i \leftarrow i-1$
    end
20. Maximum branching weight $\leftarrow \sum_{e \in BE} w(e)$

---

except for those edges incident to u_i. For any edge (x, u_i) let us denote its equivalent edge in E_{i-1} by $e = (x, y)$. This defines a vertex y on C_i and a unique edge $\tilde{e}$ in C_i which is incident to y. We also define an edge of minimum weight in C_i by e_i^0. Then the weight of each edge (x, u_i) in G_i is defined to be:

$$w(x, u_i) = w(e) - w(\tilde{e}) + w(e_i^0)$$

The motivation for this assignment will become clear in theorem 2.2. BV is modified simply by removing any vertices of BV that might be in C_i. BE is modified by removing edges of C_i and by replacing those edges with a single end-point on C_i with their equivalent edges in E_i. Notice that the latter will involve, if any, only edges incident from u_i. Edges of maximum weight into vertices of C_i are actually edges of C_i, otherwise C_i would not have been identified. When u_i is subsequently chosen in line 4, an edge incident to u_i might be added to BE at line 12 if it has, given line 7, a positive weight.

The while statement of lines 14–19 selects, in turn, those edges of each circuit $C_k, C_{k-1}, \ldots, C$ that are to be included in the branching. At line 15 G_{i-1} is reconstructed from G_i in an obvious way (in fact, a detailed algorithm might actually retain G_{i-1} when G_i is constructed from it). Also at line 11 any edges of BE with u_i as an end-point are replaced by their equivalent edges in G_{i-1}. Which edges of each C_i are added to BE depends upon whether or not BE already contains an edge incident to a vertex of C_i. The details can be seen in the conditional statement starting at line 16.

As indicated in line 20, the final set of edges in BE defines a maximum branching for G_0. We prove this in the following theorem.

Theorem 2.2. Edmond's algorithm finds a maximum branching for a weighted digraph.

Proof. We shall show that if BE contains the edges of a maximum branching for G_i then it subsequently does so for G_{i-1} in the reconstruction phase of the algorithm. If BE contains the edges of a maximum branching for the smallest constructed graph G_k, then we shall have an inductive proof that eventually BE will contain the edges of some maximum branching of G_0.

As the basis for our induction let us then consider G_k when BV contains all the vertices of V_k, as detected at line 3. Then BE contains just one edge of maximum-weight incident to each vertex of V_k provided that edge is of positive weight. Also the edges of BE are acyclic. Clearly, BE then represents a maximal branching for G_k.

Now, by the induction hypothesis, BE contains the edges of a maximum branching of G_i. Consider G_{i-1} and the corresponding set of edges BE, as redefined in the while statement starting at line 13 of the algorithm. Let E'_{i-1} be the set of edges of G_{i-1} which are incident to vertices of C_i and let E''_{i-1} be the remaining edges of G_{i-1}. We denote those edges of E'_{i-1} that are in BE by BE' and those edges of E''_{i-1} that are in BE by BE''. If BE does not represent a maximum branching of G_{i-1} then there must be a branching B with edges B' in E'_{i-1} and edges B'' in E''_{i-1} such that: either

(*a*) $w(B') > w(BE')$,

or

(*b*) $w(B'') > w(BE'')$.

In fact we shall show that BE' is a maximum branching for E'_{i-1} and that $w(B'') = w(BE'')$. Consider E'_{i-1} first.

For every vertex of C_i the incoming edge of maximum weight, and it will be of positive weight, is an edge of C_i. Otherwise C_i would not have been identified as a circuit. No maximum branching of E'_{i-1} can contain every edge of C_i. However, there is maximum branching of E'_{i-1} which includes all but one edge of C_i. This is because, from a branching excluding two or more edges of C_i, we can always obtain a branching of greater or equal weight as follows. Let v be a vertex of C_i which has no edge of C_i incident to it. Then either v has an edge e, not of C_i, incident to it or it does not. If e exists then it is replaced by another edge of greater weight, $\tilde{e}$ incident to v and in C_i, otherwise a branching of greater weight is obtained by simply including $\tilde{e}$. Now consider which of those branchings of E'_{i-1} with all but one edge of C_i is of maximum weight. Any such branching can have at most one edge of the form $e = (x, y)$, $x \notin C_i$ and $y \in C_i$, where the edge of C_i incident to y, $\tilde{e}$, would be absent. Now if such an edge e exists for a branching of E'_{i-1}, then the weight of the branching would be:

$$w(C_i)+w(e)-w(\tilde{e})$$

Let us suppose that e has been chosen to maximise this expression. Of those branchings not using an edge such as e, the one having maximum weight has the weight:

$$w(C_i)-w(e_i^0)$$

where e_i^0 is the edge of C_i with the minimum weight. Thus if

$$w(C_i)+w(e)-w(\tilde{e}) > w(C_i)-w(e_i^0)$$

that is, if

$$w(e)-w(\tilde{e})+w(e_i^0) > 0$$

then a branching of maximum weight of E'_{i-1}, which includes every edge but one of C_i, would have an edge such as e. Otherwise it would not. The left-hand side of the last inequality is precisely the weight assigned to e when C_i is shrunk to u_i in the algorithm. When the algorithm finds a branching for G_i then it includes e in BE if the inequality holds for the edge of maximum weight incident to u_i. Then line 18 assigns the edges of $C_i-\{\tilde{e}\}$ to BE' when u_i is expanded to C_i to form G_{i-1}. If the last inequality holds for no edge into u_i for G_i, then u_i becomes the root of an out-tree in BE. Line 17 of the algorithm then assigns the edges of $C_i-e_i^0$ to BE'. Thus the algorithm provides a maximum matching for E'_{i-1}.

We now show that $w(B'') = w(BE'')$. Without loss of generality we can assume that B' is of the form generated for BE' by the algorithm. If it is not then it can be converted into such a form without affecting B''. There are then two cases to consider:

(i) The branching of E'_{i-1} contains no edge outside C_i.
(ii) The branching of E'_{i-1} contains an edge $e = (x, y)$ where $x \notin C_i$ and $y \in C_i$.

By the induction hypothesis BE represents a maximum branching for G_i before u_i is expanded to C_i. In case (i) this maximum branching has a one-to-one edge correspondence with the branching produced for E''_{i-1} by the algorithm. Thus in this case $w(B'') = w(BE'')$. In case (ii) the branching of E''_{i-1} produced by the algorithm has a one-to-one edge correspondence with the maximum branching of G_i *less* the edge of this branching incident to u_i. However it is still clear that $w(B'') = w(BE'')$ because if the weight of the branching for E''_{i-1} could be increased, without including a path from a vertex of C_i to x, then the branching for G_i would not be maximal. We could construct another of greater weight from that branching with a one-to-one edge correspondence with enhanced branching of E''_{i-1} plus the edge of the original maximum branching of G_i into u_i. ■

Figure 2.4 shows an application of Edmond's algorithm. The vertices are imagined to be examined in alphabetical order, with artificial vertices being added at the tail end of the order as they are created. Starting with G_0, (*a*) shows the successive graphs G_1 and G_2 obtained in the circuit reduction stage of the algorithm. Coincidentally, for it would not generally be the case, BV and BE are empty when the processing of G_1 and G_2 starts. For G_0, G_1 and G_2 the final values of BE and BV are shown. Figure 2.4(*b*) shows the successive contents of BE as G_1 and G_0 are reconstructed. Notice that the final set of edges in BE, which define a maximum branching for G_0, is in fact a single out-tree rooted at B which does not, incidentally, include the edge of maximum weight in G_0.

Edmond's algorithm is efficient, running in $O(n|E|)$-time. The most expensive stages concern the construction and reconstruction of graphs. For each new graph we require $O(|E|)$ steps and this process is repeated no more than n times. Perhaps the only other steps of any complexity are embodied in lines 6 and 8. For each vertex v, the incoming edge of maximum weight requires $\sim d^-(v)$ comparisons. Hence line 6 requires for all vertices (even those artificially created which cannot exceed n in number) only $O(|E|)$ steps. For line 8, any circuit, if it exists, can be detected in $O(n)$-time. A newly created circuit must contain the edge denoted by e in line 8. If a circuit exists it can be detected by tracing edges in reverse direction starting at e and by visiting no more than n vertices.

If we require a minimum rather than a maximum branching, it is easy to modify Edmond's algorithm to find one. We simply replace the weight of each edge by its negative and then apply the algorithm as it has been

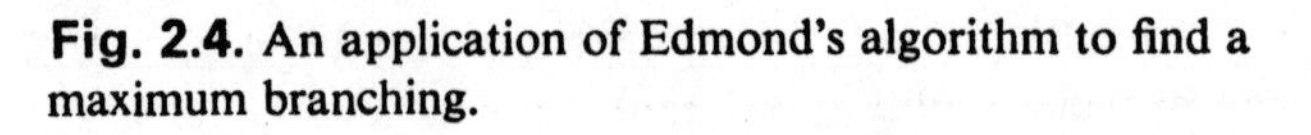

Fig. 2.4. An application of Edmond's algorithm to find a maximum branching.

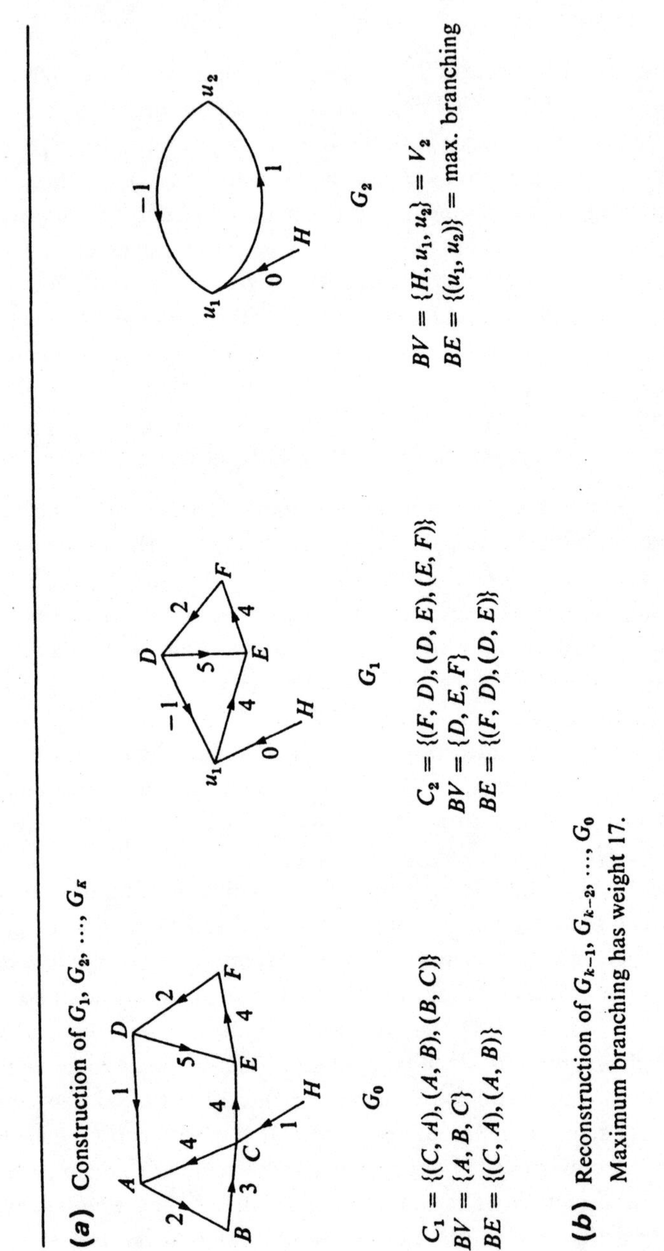

described. Obviously a maximum branching for the graph with modified edge-weights provides a minimum branching for the original graph.

We now conclude this section with a completely different type of problem.

2.1.3. Enumeration of spanning-trees

Generally a graph has a number of distinct spanning-trees. For some applications it is useful to construct spanning-trees with specific qualities. For example, there are the depth-first or the breadth-first trees which we met in chapter 1. Or we might be interested in, the much more difficult to obtain, *degree-constrained* spanning-trees in which no vertex has degree exceeding a specified value. We can describe a large variety of spanning-trees. However, we are not concerned here with their individual qualities but, rather, with the total number of trees associated with a given graph.

Before solving the general problem we prove a well-known specific result first obtained by Cayley.[5]

Theorem 2.3. The number of spanning trees of K_n is n^{n-2}.

Proof. The overall number of spanning-trees of K_n is clearly the same as the number of trees that can be constructed on n distinguished, that is, labelled vertices. Let T be a tree in which the vertices are labelled 1, 2, ..., n.

Fig. 2.5

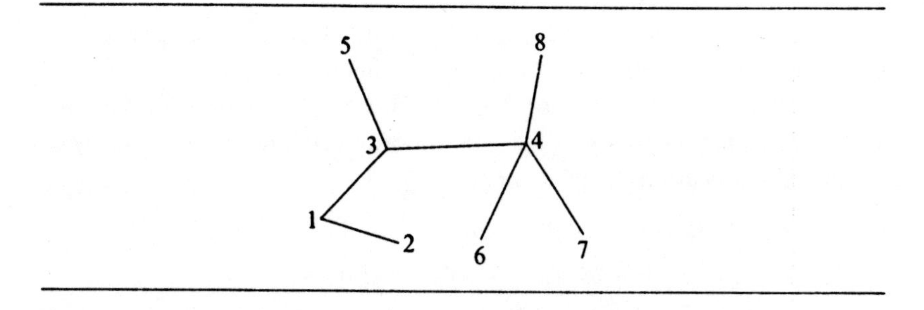

We can construct a sequence of $(n-2)$ labels, S, which uniquely encodes T as follows. In choosing the ith element of S we remove the vertex of degree one from T which has the smallest label. The ith element is then the label of the vertex remaining in T which was adjacent to the removed vertex. The process stops when only two vertices remain in T. For example, for the tree in figure 2.5 we obtain the sequence (1, 3, 3, 4, 4, 4). Notice that no vertex originally of degree one in T appears in S.

Conversely we can construct a unique tree with n vertices from a sequence S, of $(n-2)$ labels as follows. Let I be the list of labels (1, 2, ..., n).

We first look for the smallest label in I that is not in S. Let this be i_1. The edge (i_1, s_1) is then in T. We remove i_1 from I and s_1 from S and the process is repeated with the new S and the new I. Finally, S contains no elements and the last edge to be added to T is that defined by the remaining pair of labels in I.

Thus there is a one-to-one correspondence between the spanning-trees of K_n and the words of length $(n-2)$ over the alphabet $(1, 2, ..., n)$. The number of such words is n^{n-2} and so the theorem follows. ■

We come now to the general problem of counting the number of spanning trees for an arbitrary multi-graph G. This requires that we first concentrate on digraphs and counting the number of spanning out-trees rooted at a particular vertex. To this end we now introduce the so-called *Kirchoff* or *in-degree* matrix $K(G)$. The elements of K are defined as follows:

$$K(i,j) = d^{-}(v_i),\ i = j$$
$$= -k,\ i \neq j$$

where k is the number of edges from i to j. Within this definition the graph is presumed to have no self-loops. If they exist in a graph of interest, then they can be safely erased because they can make no contribution to the number of spanning trees. Figure 2.6(a) shows a digraph and its Kirchoff matrix.

Notice that the sum of the entries in any column of K is necessarily zero. We can use K to identify the set $\{g_i\}$ of subgraphs of G, in which every vertex v has $d^{-}(v) = 1$, provided in G, $d^{-}(v) \geqslant 1$. The procedure is best understood with the aid of an example. In figure 2.6(b) the determinant of K, for the graph of figure 2.6(a) is expanded into a sum of determinants, each corresponding to some g_i. This procedure is always possible by a continued application of the identity:

$$\det(c_1, c_2, ..., (c_j + c_j'), ..., c_n)$$
$$= \det(c_1, c_2, ..., c_j, ..., c_n) + \det(c_1, c_2, ..., c_j', ..., c_n)$$

where each c_i is a column of n elements. Subsequent applications are used to reduce the value of a diagonal element which is greater than one and to produce two determinants, each of which has the sum of the elements in any column equal to zero. Thus each is of a Kirchoff matrix for some graph. The expansion stops when every diagonal element of the determinants produced are not greater than one. We then have

$$\det(K(G)) = \sum_i \det(K(g_i))$$

where $K(g_i)$ is the in-degree matrix of g_i. In our example g_i is drawn below its associated $\det(K(g_i))$ in figure 2.6(b).

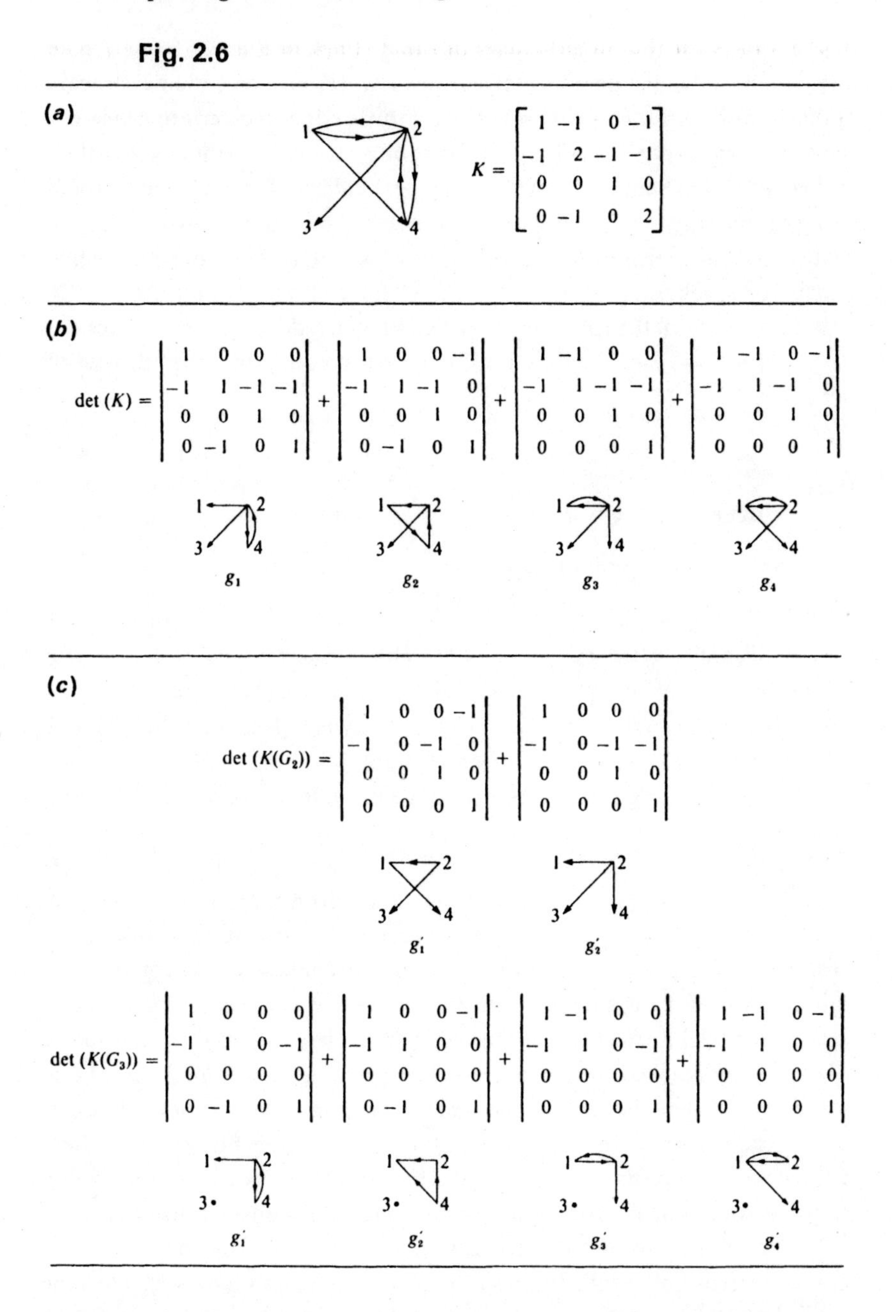

In the expansion of figure 2.6(*b*) each g_i corresponds to a subgraph of G in which $d^-(v) = 1$, if in G, $d^-(v) \geqslant 1$. Clearly, every such subgraph of G is represented precisely once in this expansion. Consider the spanning out-trees rooted at a particular vertex r of G and let T be such a tree. If $d^-(r) > 1$

then T appears within exactly $d^-(r)$ of the g_i. Each of these g_i being T plus one possible edge incident to r. If, however, $d^-(r) = 0$ or 1 then T appears in one g_i only. If we let G_r denote G with those edges incident to r deleted, then we can expand $\det(K(G_r))$ according to the method prescribed earlier. In this expansion, however, any particular out-tree rooted at r will be represented by exactly one term. In figure 2.6(c) we see two examples. Notice that each term in these expansions is the determinant of an in-degree matrix for a subgraph g_i' of G in which $d^-(v) \leqslant 1$, $v \neq r$, and $d^-(r) = 0$.

We now require the following theorem in which $\det(K_{rr}(G))$ denotes the minor resulting from the deletion of the rth column and the rth row of $\det(K(G))$.

Theorem 2.4. If g is a finite digraph such that for each vertex v, $d^-(v) \leqslant 1$, then

$$\det(K_{rr}(g)) = 1 \text{ if } g \text{ contains a spanning out-tree rooted at } r$$
$$= 0 \text{ otherwise}$$

Proof. Suppose that g contains a spanning out-tree T rooted at r and that its vertices are labelled 1, 2, ..., n. Then either $g = T$ or $g = T+e$ where e is an edge incident to r. We can relabel the vertices according to a breadth-first order of traversing T, visiting r first. Then $r = 1$ so that $K(1, 1) \leqslant 1$ and for $i > 1$, $K(i, i) = 1$. Also if $i \neq 1$ and $i > j$ then $K(i, j) = 0$. Thus $K_{11}(g)$ is an upper-right-triangular matrix with unit diagonal elements, and so $\det(K_{11}(g)) = 1$.

Now suppose that g does not contain a spanning out-tree rooted at r. If any vertex $v \neq r$ has $d^-(v) = 0$ then the corresponding column of K consists of zeros only, so that $\det(K_{rr}(g)) = 0$. Suppose then that every vertex $v \neq r$ has $d^-(v) = 1$. Since g does not contain a tree rooted at r, then it must contain a circuit which excludes r as follows. Trace the edge into $v_i \neq r$ backwards to v_j and the edge into v_j backwards to v_k and so on. If r is finally reached in this process then $v_i, v_j, v_k, \ldots$ belong to a subtree rooted at r. Otherwise the process ends up tracing a circuit not including r. If r is reached we repeat the process starting at vertices not in the subtree. Clearly we can accommodate every vertex, without constructing a spanning out-tree rooted at r, only by completing a circuit. Consider now the set of columns of K corresponding to vertices on a circuit. Any row of this set contains zeros only or it contains a single (-1) and a single $(+1)$. Thus the sum of these columns is a column of zeros. It follows that $\det(K_{rr}(g)) = 0$. ■

The g_i' (and incidentally the g_i) defined earlier satisfy theorem 2.4. We note that $K_{rr}(G)$ is identical to $K(G_r)$ except that the (rr)th element is unity instead of zero. There is obviously a one-to-one correspondence between the terms of det $(K_{rr}(G))$ and of det $(K(G_r))$ if each is expanded according to the prescription given earlier. In fact:

$$\det(K_{rr}(G)) = \sum_i \det(K_{rr}(g'_i))$$

and if we apply theorem 2.4 to each term of the above sum we immediately obtain the following theorem.

Theorem 2.5. The number of spanning out-trees rooted at r in a finite digraph G is equal to det $(K_{rr}(G))$.

Figure 2.7 illustrates an efficient algorithm based on theorem 2.5 to calculate the number of spanning out-trees rooted at r. Now, K is an $n \times n$ Kirchoff matrix and line 1 of the algorithm assigns (or inputs) the elements of K_{rr} to an $(n-1)\times(n-1)$ determinant A. This can clearly be done in $O(n^2)$ steps. Lines 2–5 use a Gaussian method to convert A into an upper triangular form. In other words, a series of weighted row subtractions reduces each element $A(i, j)$ of A, for which $i > j$, to zero. This is achieved in $O(n^3)$ steps. The determinant is then evaluated by a diagonal expansion in line 7 using $O(n)$ steps, the result being assigned to $DKRR$. Overall, therefore, figure 2.7 illustrates an $O(n^3)$-algorithm.

Fig. 2.7. An algorithm to find the number of spanning out-trees rooted at r in a digraph with in-degree matrix K, or the number of spanning trees of an undirected graph with degree matrix K.

```
1. A ← K_rr
2. for k = 2 to (n−1) do
3.         for i = k to (n−1) do
4.                 for j = 1 to (n−1) do
5. A(i,j) ← A(i,j)−(A(i, (k−1))/A((k−1), (k−1))).A((k−1), j)
6. DKRR ← A(1, 1)
7. for i = 2 to (n−1) do DKRR ← DKRR.A(i, i)
```

Having resolved the problem of counting the number of spanning out-trees rooted at a given vertex in a digraph, we can very quickly see a solution to the problem of counting the spanning-trees of an undirected graph. To do this we note, given an undirected graph G, that we can construct a digraph G' by replacing each edge (u, v) by two directed edges (u, v) and (v, u). Then for every spanning-tree of G there corresponds a spanning out-tree in G' *rooted at a particular vertex*, and vice-versa. We define the *degree* matrix of G to be identical to the in-degree matrix of G', and we denote it also by K. We therefore have the following theorem.

Theorem 2.6. The number of spanning-trees in a finite undirected graph is equal to any one of the minors det $(K_{rr}(G))$ for $1 \leqslant r \leqslant n$.

The theorem embodies the obvious conclusion that the number of spanning trees cannot depend on the choice of r. As the caption to figure 2.7 implies, we can obviously use that $O(n^3)$-algorithm to count the number of spanning-trees of an undirected graph as well as to count the spanning out-trees rooted at a particular vertex in a digraph.

2.2 Circuits, cut-sets and connectivity

In this section we demonstrate the importance of spanning-trees with respect to the circuit space and the so-called cut-set space of a graph. We shall also be concerned with the separability of a graph by generalising the notions of articulation point and cut-edge which were introduced in chapter 1.

It is convenient here to extend our definitions. A *co-tree* of a graph $G = (V, E)$ with respect to a spanning-tree $T = (V, E')$ is the set of edges $(E-E')$. If G has n vertices then any co-tree, if one exists, has $|E|-(n-1)$ edges. Any edge of a co-tree is called a *chord* of the spanning-tree. We need also to define the operation of *ring-sum*. The ring-sum of two graphs $G_1 = (V_1, E_1)$ and $G_2 = (V_2, E_2)$, which we write $G_1 \oplus G_2$, is the graph $((V_1 \cup V_2), ((E_1 \cup E_2)-(E_1 \cap E_2))$. In other words the edge-set of $G_1 \oplus G_2$ consists of those edges which are either in G_1 or are in G_2 but which are not in both. It is easy to see that the operation of ring-sum is both commutative and associative. That is, that:

$$G_1 \oplus G_2 = G_2 \oplus G_1$$

and that

$$(G_1 \oplus G_2) \oplus G_3 = G_1 \oplus (G_2 \oplus G_3)$$

2.2.1. *Fundamental circuits of a graph*

From theorem 1.2 we see that the addition of a chord to a spanning-tree of a graph creates precisely one circuit. In a graph the collection of these circuits with respect to a particular spanning-tree is called a set of *fundamental circuits*. As we shall see, any arbitrary circuit of the graph may be expressed as a linear combination of the fundamental circuits using the operation of ring-sum. In other words, the fundamental circuits form a *basis* for the circuit space.

Figure 2.8 shows, for the graph illustrated there, a spanning-tree T, the corresponding set of fundamental circuits and some other circuits expressed

Fig. 2.8. Some circuits of G expressed as linear combinations of the fundamental circuits of G with respect to T.

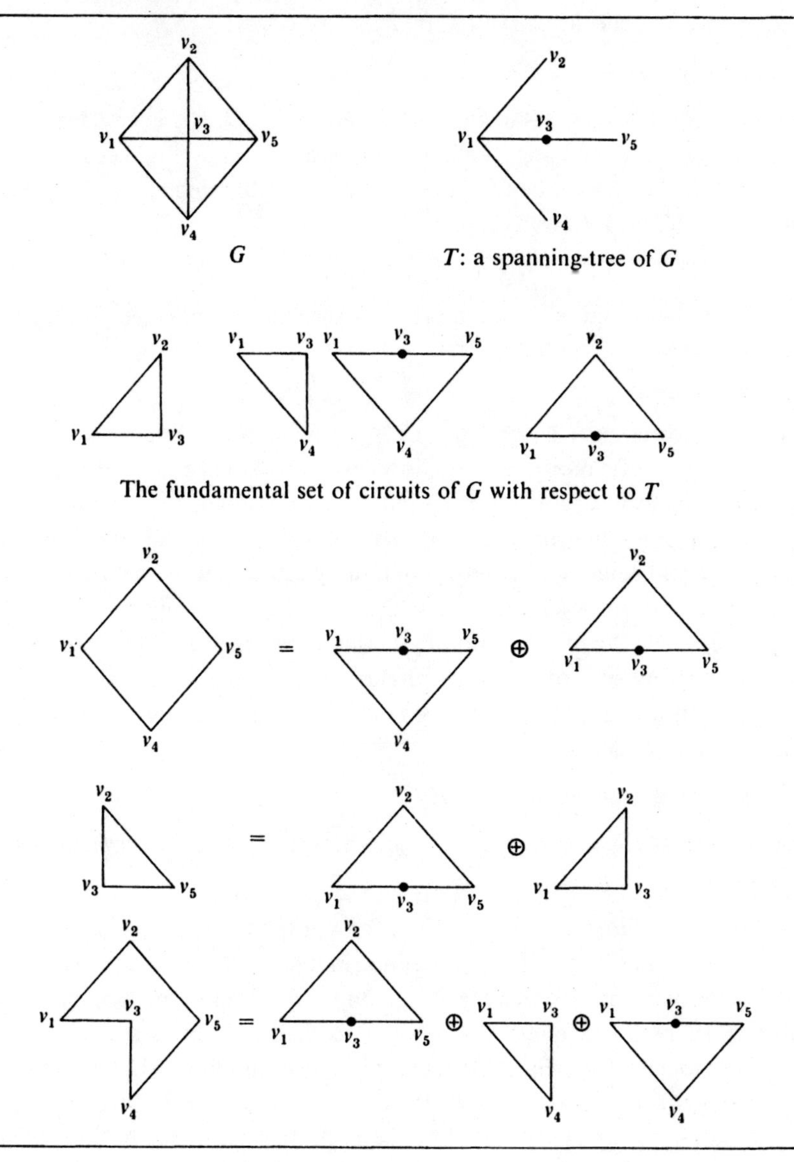

as linear combinations of these. In general then, we have the following theorem.

Theorem 2.7. A set of fundamental circuits, with respect to some spanning-tree of a graph G, forms a basis for the circuit space of G.

Proof. We first show that any circuit can be expressed as a linear combination of the fundamental circuits F with respect to some spanning-tree T. We denote an arbitrary circuit C by its set of edges:

$$C = \{e_1, e_2, \ldots, e_i, e_{i+1}, \ldots, e_j\}$$

where e_k for $1 \leqslant k \leqslant i$ is a chord of T and for $i < k \leqslant j$ is an edge of T. F contains precisely one fundamental circuit containing each e_k for $1 \leqslant k \leqslant i$. We denote the fundamental circuit containing e_k by $C(e_k)$. We now define C' as follows:

$$C' = C(e_1) \oplus C(e_2) \oplus \ldots \oplus C(e_k)$$

and show that C and C' contain precisely the same set of edges. If they do not then:

$$C \oplus C' \neq \varnothing$$

For any two circuits C_1 and C_2, $C_1 \oplus C_2$ must be a circuit or an edge disjoint union of circuits. Thus C' is a circuit or an edge disjoint union of circuits and so is $C \oplus C'$. But C and C' each contain the set of chords $e_1, e_2, \ldots, e_k$ of T and no other chords. Thus $C \oplus C'$ could only contain edges of T and could not therefore contain a circuit. Thus we have a contradiction so that our assumption that $C \neq C'$ must be wrong.

We complete the proof by noting that no member of F can be expressed as a linear ring-sum of the other circuits of F. This follows immediately from the observation that each chord of T is contained in one and only one fundamental circuit. ■

We have an immediate corollary:

Corollary 2.1. The circuit space for a graph with $|E|$ edges and n vertices has dimension $(|E|-n+1)$.

A set of fundamental circuits, *FCS*, for a graph G can easily be found in polynomial time. The algorithm outlined in figure 2.9 for example operates in $O(n^3)$-time. Line 1 states that a spanning-tree T and the corresponding co-tree CT of G are found first. We saw in chapter 1 that a spanning-tree can be found in $O(\max(n, |E|))$-time. It is easy to modify such an algorithm so that when an edge is found not to be required for T, then it is not discarded but is added to CT. What is more, this can easily be done at no cost to the order of the complexity. Thus line 1 can be achieved in $O(\max(n, |E|))$-time.

For each edge $e_i \in CT$ the body of the for statement in lines 4–6 finds one fundamental circuit and adds it to the set *FCS*. The search for the path in T which makes a circuit with e_i in line 4 can be achieved in $O(n)$ steps as follows. Vertex v_i in T is labelled one. Then starting at v_i and using a breadth-first scan of T, each vertex v is labelled $(L+1)$ where L is the label of the father of v. The process stops when v_i' acquires a label. The path from v_i' to v_i is then easily traced by starting at v_i' and proceeding so that at each step the next vertex visited has a label which is numerically one less than that for the current vertex. In this subalgorithm the initial labelling can clearly be achieved in O(n) steps. The total time spent scanning edges at individual vertices when tracing the path is O(n), at worst each edge is scanned twice except for the initial and final edges and T has $(n-1)$ edges. The path itself is the accumulation of at most $(n-1)$ edges and so the whole path finding process requires no more than O(n)-time.

Fig. 2.9

1. Find a spanning-tree T and the corresponding co-tree CT of G.
2. $FCS \leftarrow \emptyset$
3. **for all** $e_i = (v_i, v_i') \in CT$ **do**
   begin
4. find the path from v_i to v_i' in T and denote it by P_i
5. $C_i \leftarrow P_i \cup \{e_i\}$
6. $FCS \leftarrow FCS \cup C_i$
   end

The number of edges in CT is O($|E|$), that is, O(n^2) generally or O(n) for a sparse graph. Thus the body of the for statement is executed O($|E|$)-times so that lines 3–6, which essentially determine the overall complexity of the algorithm, can be executed in O(n^2)-time for a sparse graph and, at worst, in O(n^3)-time.

Kirchoff[6] was an early developer of the theory of trees, in his case in connection with electrical circuits. A well-known consequence of theorem 2.7 and its corollary concerns Kirchoff's voltage law. That is, that the net voltage drop around any cycle of an electrical circuit is zero. This law is generally used to obtain a set of simultaneous equations in the unknown voltage drops across individual components of the network. One equation is obtainable from each circuit of the network. Theorem 2.7 and its corollary tell us which circuits of the underlying graph of the network, and how many of them, provide a linearly independent set of equations.

2.2.2. Fundamental cut-sets of a graph

A *cut-set* of a connected graph, or component, is a set of edges whose removal would disconnect the graph or component. As a part of the definition no proper subset of a cut-set will cause disconnection. A consequence of this is that a cut-set produces exactly two components.

It is sometimes useful to denote a cut-set by the partition of vertices that it induces. If V denotes the vertex-set of G and if P is the subset of vertices in one component of G induced by the cut-set, then the cut-set can be specified by $(P, \bar{P})$ where $\bar{P} = V - P$.

As we shall see in chapter 4, cut-sets play an important rôle in the study of transport networks. Also another practical application concerns the vulnerability of communicating systems with respect to failure. In a graph whose edges represent the lines of communication in such a system, the weakest link is the cut-set of smallest size.

In the previous section we defined a basis for the circuits of a graph in terms of fundamental circuits. We shall similarly define a set of *fundamental cut-sets*. Again, the idea of a spanning-tree plays an important rôle here. Let T be such a spanning-tree of the connected graph G. Any edge of T defines a partition of the vertices of G since its removal disconnects T into two components. There will be a corresponding cut-set of G producing the same partition of vertices. This cut-set contains precisely *one* edge and a number of chords of T. Such a cut-set is called a *fundamental cut-set* of G with respect to T. Figure 2.10 shows, for the graph of that diagram, a spanning-tree, a corresponding set of fundamental cut-sets and some other cut-sets expressed as linear ring-sums of fundamental cut-sets. In general we have theorem 2.8.

Fig. 2.10.

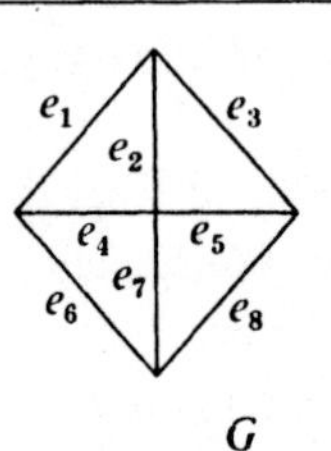

G

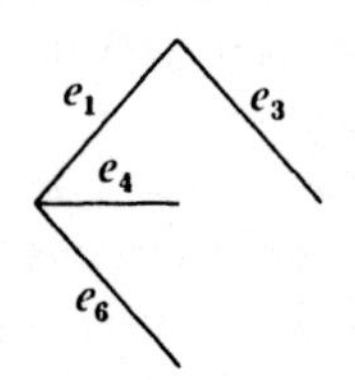

T: a spanning-tree of G

$$C_1 = \{e_1, e_2, e_5, e_8\}$$
$$C_2 = \{e_4, e_2, e_5, e_7\}$$
$$C_3 = \{e_6, e_7, e_8\}$$
$$C_4 = \{e_3, e_5, e_8\}$$

The set of fundamental cut-sets of G with respect to T

$$\{e_3, e_5, e_6, e_7\} = C_3 \oplus C_4$$
$$\{e_1, e_4, e_6\} = C_1 \oplus C_2 \oplus C_3$$
$$\{e_1, e_2, e_5, e_8\} = C_1 \oplus C_2 \oplus C_3 \oplus C_4$$

Some cut-sets of G expressed as linear combinations of the fundamental cut-sets of G with respect to T.

Theorem 2.8. The fundamental cut-sets with respect to some spanning-tree form a basis for the cut-sets of a graph.

Proof. This is entirely analogous to that for theorem 2.7 and so we omit the details. The main function of the proof, given an arbitrary cut-set CS:

$$CS = e_1, e_2, \ldots, e_i, e_{i+1}, \ldots, e_j$$

where e_k for $1 \leqslant k \leqslant i$ are edges of the spanning-tree and e_k for $i < k \leqslant j$ are chords, is to show that CS is identical to CS':

$$CS' = CS(e_1) \oplus CS(e_2) \oplus \ldots \oplus CS(e_i)$$

where $CS(e)$ is the fundamental cut-set associated with the edge e of T.

We proceed exactly as in the proof of theorem 2.7, the only difficulty here is that it may not be immediately obvious that the ring-sum of two cut-sets is a cut-set or an edge disjoint union of cut-sets. We can see this informally as follows. Let $C_1 = (V_1, \bar{V}_1)$ and $C_2 = (V_2, \bar{V}_2)$ be cut-sets of a graph G. If the edges of both C_1 and C_2 are removed from G, then the vertices are partitioned into four subsets $(V_1 \cap V_2)$, $(V_1 \cap \bar{V}_2)$ $(\bar{V}_1 \cap V_2)$ and $(\bar{V}_1 \cap \bar{V}_2)$ such that no remaining edge of G connects vertices in different subsets. The ring sum of C_1 and C_2 consists of those edges in C_1 and those in C_2 but not those in *both* C_1 and C_2. Those edges common to both C_1 and C_2 can only connect vertices in $(V_1 \cap V_2)$ to vertices in $(\bar{V}_1 \cap \bar{V}_2)$ *or* vertices in $(\bar{V}_1 \cap V_2)$ to vertices in $(\bar{V}_1 \cap V_2)$. Thus if the edges of $C_1 \oplus C_2$ are removed from G then there is a partitioning of the vertices into $(V_1 \cap V_2) \cup (\bar{V}_1 \cap \bar{V}_2)$ and $(V_1 \cap \bar{V}_2) \cup (\bar{V}_1 \cap V_2)$. If each of these subsets induces a connected subgraph then $C_1 \oplus C_2$ is a cut-set. Otherwise it is an edge disjoint union of cut-sets. Figure 2.11 illustrates this. ■

Fig. 2.11. Illustrating, with reference to figure 2.10, that the ring-sum of two arbitrary cut-sets is either a cut-set or the edge-disjoint union of cut-sets.

$$\{e_1, e_4, e_6\} \oplus \{e_2, e_3, e_4, e_6\} = \{e_1, e_2, e_3\}$$
$$\{e_1, e_2, e_7, e_8\} \oplus \{e_3, e_5, e_6, e_7\} = \{e_1, e_3, e_4, e_5, e_6, e_8\}$$
$$= \{e_1, e_4, e_6\} \cup \{e_3, e_5, e_8\}$$

We have the following corollary.

Corollary 2.2. The cut-set space for a graph with n vertices has dimension $(n-1)$.

As for the case of fundamental circuits, it is easy to construct a polynomial time algorithm to find a set of fundamental cut-sets. See, for example, exercise 2.8.

2.2.3. Connectivity

Circuits and cut-sets are aspects of the connectedness of a graph. It is natural therefore that we should generalise here those two basic definitions of separability, cut-edge and articulation point, which were introduced in chapter 1. This leads naturally to consideration of the number of edge disjoint paths between any two distinct vertices.

It is clear that removing any vertex from a tree will disconnect it. On the other hand, the removal of any vertex or subset of vertices from a complete graph will not disconnect it. Trees and complete graphs represent the two extreme cases of *vertex-connectivity*, or simply *connectivity* of a graph. For an arbitrary graph G, we define its connectivity, written $K_v(G)$ or simply K_v, to be the minimum number of vertices whose removal will disconnect G. Also we say that G is *h-connected* for any positive integer h satisfying $h \leqslant K_v(G)$. Any subset of vertices whose removal will disconnect G is called a *vertex-cut*.

Similarly we define the *edge-connectivity*, $K_e(G)$ or K_e, for the connected graph G to be the size of the smallest cut-set of G. G is said to be *h-edge-connected* for any positive integer h satisfying $h \leqslant K_e(G)$.

We denote the smallest degree of any vertex in a graph by δ. Since the set of edges incident with any vertex forms a cut-set, we have that $\delta \geqslant K_e(G)$. Also $K_v(G)$ cannot exceed $K_e(G)$. We can see this informally by recognising that a vertex-cut is obtainable by removing an end-point from each edge of a minimum cut-set. For convenience we define $K_v(K_n)$ to be $(n-1)$. We then have the following theorem:

Theorem 2.9. For any connected graph G:

$$K_v(G) \leqslant K_e(G) \leqslant \delta$$

We now describe a theorem for 2-connected graphs before stating its generalisation to graphs which are h-connected.

Theorem 2.10. A graph G with at least three vertices is a block if and only if two vertices are connected by at least two edge disjoint paths.

Proof. If any two vertices of G are connected by at least two edge disjoint paths then G is connected and cannot contain an articulation point. Therefore G must be a block.

On the other hand, suppose that G is 2-connected. Let $l(u, v)$ be the path length from u to v. We prove, by induction on $l(u, v)$, that there are two edge disjoint paths from u to v. If $l(u, v) = 1$, then (u, v) cannot be a cut-edge if G is a block. Then $G-(u, v)$ is connected and so G contains a path from u to v which does not utilise (u, v). We thus have a basis for our induction. Let us assume then that there are two edge disjoint paths between u and v for $l(u, v) < L$. Now suppose that $l(u, v) = L$ and let P be a path of length L from u to v and let w be the vertex adjacent to u on P. By the induction hypothesis v and w are connected by two edge disjoint paths. Without loss of generality, we can take one to be $P-(u, w)$ and we denote the other by Q. Since G is a block, $G-w$ must be connected. Let R be the path from u to v not including w, and let x be the first vertex common to R and Q which is encountered by following R from u. It is possible that $x = u$. Clearly, there are two edge disjoint paths from u to v. One is

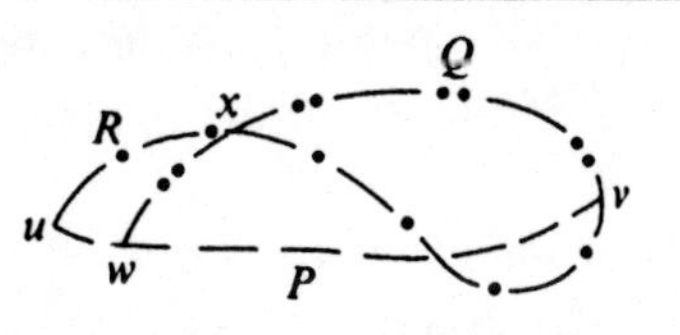

P and the other is that portion of R from u to x plus that portion of Q from x to v. ■

We have two corollaries:

Corollary 2.3. In a block with at least three vertices any two vertices lie on a common cycle.

Corollary 2.4. In a block with at least three vertices any two edges lie on a common cycle.

Proof. Let G be a block with at least three vertices. Let any two of its edges by (u_1, v_1) and (u_2, v_2). From G we construct G', a block with at least five vertices by adding two vertices, w_1 and w_2, both of degree 2. This is done by replacing (u_1, v_1) with the edges (u_1, w_1) and (w_1, v_1) and by replacing (u_2, v_2) with (u_2, w_2) and (w_2, v_2). From corollary 2.3, w_1 and w_2 of G' lie on a common cycle. It follows that (u_1, v_1) and (u_2, v_2) of G lie on a common cycle. ■

In chapter 4 we prove a well-known theorem, due to Menger,[7] which is a generalisation of theorem 2.10. This is that a graph with at least $(h+1)$ vertices is h-connected if and only if two distinct vertices are connected by at least h edge disjoint paths. There are, of course, also generalisations of the above corollaries.

We shall delay until chapter 4 presentation of algorithms to determine $K_v(G)$ and $K_e(G)$ for an arbitrary graph G.

2.3 Summary and references

Our concern in this chapter has been with subgraphs that are trees. We described Prim's[1] algorithm which finds optimal weight trees for undirected graphs and Edmond's[8] algorithm for optimal branchings. For alternative insight into theorem 2.2 see Karp.[9] Edmond's algorithm may also be used to find optimal weight *spanning* out-trees (or in-trees) if they exist. See, for example, exercise 2.10. We shall utilise both the idea of a spanning-tree and the enumeration of spanning-trees in chapter 3. Kirchoff[6] first described, amongst other material, a matricial method for counting spanning-trees. Excercise 2.13 describes another interesting, although inefficient, method to count the spanning-trees of an undirected graph.

In the second half of the chapter we showed how spanning-trees give particular insight into the structure of a graph as viewed from its circuit and cut-set spaces. The intimate connection between cut-sets and circuits will be further pursued in chapter 4.

Theorem 2.10 is due to Whitney[10] and its generalisations due to Menger[7] will be proved in chapter 4. We shall also in that chapter describe algorithms to determine the edge- and vertex-connectivities of a graph.

Those readers interested in pursuing the specific application areas of trees mentioned in the first paragraph of the chapter might refer to Hopcroft & Ullman,[12] Knuth[3] and Bell[14]. For general reading on trees the following are recommended. Chapters 2, 3 and 12 of Bondy & Murty,[18] chapters 2–4 of Deo,[15] chapters 1 and 2 of Busacker & Saaty,[16] chapters 4 and 15 of Harary[17] and chapters 12, 13 and 16 of Berge.[11]

[1] Prim, R. C. 'Shortest connection networks and some generalisations', *Bell System Tech. J.*, **36**, 1389–401 (1957).
[2] Kruskal, J. B. Jnr. 'On the shortest spanning sub-tree and the travelling salesman problem'. *Proc. Amer. Maths Soc.*, **7**, 48–50 (1956).
[3] Cheriton, D. & Tarjan, R. E. 'Finding minimum spanning-trees', *SIAM J. on Comput.*, **5** (4), 724–42 (1976).
[4] Chang, S. K. 'The generation of minimal trees in a Steiner topology', *JACM.*, **19** (4), 699–711 (1972).
[5] Cayley, A. 'On the theory of analytical forms called trees', *Phil. Mag.*, **13**, 172–6 (1857).
[6] Kirchoff, G. 'Über die Ausflösung der Gleichungen auf welche man bei der Untersuchungen der Linearen Verteilung Galvanisher Ströme geführt wird', *Poggendorf Ann. Physik*, **72**, 497–508 (1947).
[7] Menger, K. 'Zur allgemeinen Kurventheorie', *Fund. Math.*, **10**, 96–115 (1927).
[8] Edmonds, J. 'Optimum branchings', *J. of Res. of the Nat. Bureau of Standards., Sci. Sinica*, **14**, 1396–400 (1965).
[9] Karp, R. M. 'A simple derivation of Edmonds algorithm for optimum branchings', *Networks*, **1**, 265–72 (1972).

[10] Whitney, H. 'Nonseparable and planar graphs', *Trans. Amer. Math. Soc.*, **34**, 339–62 (1932).
[11] Berge, C. *The Theory of Graphs and its Applications.* John Wiley and Sons (1962).
[12] Hopcroft, J. E. & Ullman, J. D. *Introduction to Automata Theory, Languages and Computation.* Addison-Wesley (1979).
[13] Knuth, D. E. *The Art of Computer Programming.*, vol. 3, *Sorting and Searching.* Addison-Wesley (1973).
[14] Bell, A. G. *Games Playing with Computers.* George Allen and Unwin, London (1972).
[15] Deo, N. *Graph Theory with Applications to Engineering and Computer Science.* Prentice-Hall (1974).
[16] Busacker, R. G. & Saaty, T. L. *Finite Graphs and Networks.* McGraw Hill (1965).
[17] Harary, F. *Graph Theory.* Addison-Wesley (1969).
[18] Bondy, J. A. & Murty, U. S. R. *Graph Theory with Applications.* The Macmillan Press (1976).

EXERCISES

2.1. Given a specific edge e of an undirected graph G, how would you construct a spanning-tree of G which contains e? How can a graph be constructed given the set of all its spanning-trees?

2.2. Use theorem 2.5 to derive Cayley's theorem that the number of spanning-trees of K_n is n^{n-2}.

2.3. Theorem 2.5 provides a method to count the number of spanning out-trees rooted at a particular vertex of a digraph. Contrive a similar method to count the number of spanning in-trees.

2.4. The following algorithm due to Kruskal[2] finds a minimum-weight spanning-tree, MWT, of a weighted undirected graph $G = (V, E)$. Show that it operates in polynomial time.

1. Relabel the elements of E so that
   if $w(e_i) > w(e_j)$ **then** $i > j$
2. $MWT \leftarrow \varnothing$
3. **for** $i = 1$ **to** $|E|$ **do**
   if $MWT \cup \{e_i\}$ is acyclic **then**
   $MWT \leftarrow MWT \cup \{e_i\}$

(Also see exercise 2.6.)

2.5. Given a weighted undirected graph $G = (V, E)$, let V' be a proper subset of its vertices. Also let e denote the edge of smallest weight with one end in V' and the other in $(V - V')$. Show that there exists a minimum-weight spanning-tree of G which contains e.

(Let T be a minimum-weight spanning-tree of G. If T does not contain e, then $(T+e)$ contains a circuit, C. Let $e' \neq e$ be an edge of C. T' then defines a tree:

$T' = T+e-e'$

Clearly,

$w(T') \leqslant w(T)$)

2.6. Prove that Kruskal's algorithm (exercise 2.4) finds a minimum-weight spanning-tree.
(Let T_K be a spanning-tree constructed according to Kruskal's algorithm and let T_M be a minimum-weight spanning-tree. We assume that the edges are ordered, as in exercise 2.4, according to non-decreasing weight. Change T_M into T_K by a series of edge replacements each as follows. If s be the smallest value of i such that $e_i \in T_K$ but $e_i \notin T_M$, construct $T'_M = T_M + e_s - e'_s$ such that $e'_s \notin T_K$ and e'_s is an edge of the circuit in $(T_M + e_s)$. Show that T'_M must also be a minimum-weight spanning-tree of the graph.)

2.7. Let T be a particular spanning-tree of an undirected graph G. By c we denote a chord and by e an edge of T. Justify the following statements:
(*a*) If S is a fundamental circuit defined by T and c, then c appears in each fundamental cut-set defined by an edge in $(S - \{c\})$ and in no others.
(*b*) If K is a fundamental cut-set defined by T and e, then e appears in each fundamental circuit defined by a chord in K and in no others.

2.8. Construct a polynomial time algorithm to find a set of fundamental cut-sets for some undirected graph $G = (V, E)$.
(An $O(n^3)$-algorithm may be constructed as follows. First find a spanning-tree T of G. Then for each edge $e \in T$ determine the two blocks B_1 and B_2 of $(T - \{e\})$. The fundamental cut-set associated with e, $FCS(e)$ is then given by:

$FCS(e) \leftarrow \varnothing$
for all $v_i \in B_1$ **do**
  for all $v_j \in B_2$ **do**
    if $(v_i, v_j) \in E$ **then** $FCS(e) \leftarrow FCS(e) \cup \{(v_i, v_j)\}$)

2.9. Construct a counterexample to show that the following 'algorithm' does not always construct a maximum branching, MB, of a weighted digraph $G = (V, E)$.

1. Relabel the elements of E so that
   if $w(e_i) > w(e_j)$ **then** $i > j$.
2. $MB \leftarrow \varnothing$
3. **for** $i = 1$ **to** $|E|$ **do**
   if $w(e_i) > 0$ **and** $MB \cup \{e_i\}$ is acyclic **then**
     $MB \leftarrow MB \cup \{e_i\}$.

2.10. Given a weighted directed graph, modify Edmond's algorithm to find a maximum-weight *spanning out-tree* if one exists.
(Consider the effect of adding a constant positive weight to the existing weight of each edge of the graph.)

2.11. Justify the following statements:
(*a*) If G is a simple graph then

$$K_v(G) \leqslant \frac{2 \cdot |E|}{n}.$$

(*b*) If G is simple and 3-regular then

$$K_v(G) = K_e(G).$$

(Note theorem 2.9.)

2.12. The *connector problem* may be modified by insisting that certain pairs of locations be *directly* linked. Modify Prim's algorithm to accommodate this.

(From the original weighted graph G of the problem, construct a new graph G^*, by contracting every edge that *must* appear in the solution. Consider applying Prim's algorithm to G^*.)

2.13. $N(G)$ denotes the number of spanning-trees of the undirected graph $G = (V, E)$. Show that the following recursive formula holds:

$$N(G) = N(G-e)+N(G \circ e)$$

where $e \in G$ and $G \circ e$ means the graph obtained from G by contracting the edge e.

Show that the implied algorithm for calculating $N(G)$ has exponential time-complexity.

($N(G-e)$ is the number of spanning-trees of G not using e. To every spanning-tree of $G \circ e$, there corresponds exactly one spanning-tree of G that uses e.)

2.14. Consider again Kruskal's algorithm of exercise 2.4. We provide an outline of it once more but in the following form:

```
1. Construct a priority queue based on the edge-weights
2. MWT ← ∅
3. Assign a 'component' number L(v) to each vertex v
4. for i = 1 to |E| do
        begin
5.      Remove the edge (u, v) of minimum weight from the
            priority queue
6.      if L(u) ≠ L(v) then
                  begin
7.                Unite (C(u), C(v))
8.                MWT ← MWT ∪ {(u, v)}
                  end
        end
```

We described priority queues in exercise 1.16. In the course of constructing MWT, the set of edges which eventually becomes the minimum-weight spanning-tree, MWT contains a set of connected components each of which is a tree (initially just a single vertex). Each vertex in the same component has the same 'component' number $L(v)$ which is different for vertices in different components. The purpose of line 7 is to make the vertices in the component $C(u)$ which contains u and the vertices in the component $C(v)$ which contains v, have the same component number. This could be the component number of v or of u. Notice the condition of line 6. $L(u) \neq L(v)$ shows that u is not in the same component as v. If $L(u) = L(v)$ then $MWT \cup \{(u, v)\}$ would contain a circuit.

Fill in the implementation details which are missing from the algorithm outline and which will ensure that it runs in $O(|E| \log |E|)$-time. Notice that the use of a priority queue (exercise 1.16(*a*)) ensures that, for all iterations, line 5 requires only $O(|E| \log |E|)$-time. The remaining difficulty concerns line 7. We can see, in principle, that the total time spent executing all iterations of this instruction can be contained within the specified limit as follows. Two components $C(u)$ and $C(v)$ are united into a single component by changing the component number of the vertices in the smaller component to the component number of the *larger* component. Consider the total number of vertex component number changes there will be. Each vertex, after changing its component number will belong to a component which is at least twice as big as its component before the change. Hence, if a vertex has its component number changed i times, it belongs to a component containing at least 2^i vertices. This cannot exceed n, so that the maximum value of i is $\log n$. For all vertices, the total number of component number changes is therefore $O(n \log n)$. Your implementation details should ensure that this is the case.

Obviously an $O(|E| \log |E|)$-algorithm will be preferable to an $O(n^2)$-algorithm (described in the text) for graphs with relatively few edges.

3

Planar graphs

Our primary interest in this chapter is to determine what graphs can be arranged on a plane surface such that no two edges cross and such that no two end-points coincide. Further, we describe one algorithm to show that for an arbitrary graph efficient algorithms exist to determine whether or not it falls within this category.

This question of the *planarity* of a graph, apart from its theoretical interest, has a number of practical applications. For example, in the layout of electronic circuits, does a planar representation of a given circuit exist? If not what is the minimum number of planar graphs whose union is a representation of the circuit?

3.1 Basic properties of planar graphs

In this first section we outline some basic properties of planar graphs. As has already been stated, a graph is planar if it can be drawn on a plane surface with no two edges intersecting. More precisely, a graph G is *planar* if it is isomorphic to a graph G' such that the vertices and edges of G' are contained in the same plane and such that at most one vertex occupies or at most one edge passes through any point of the plane. G' is said to be *embedded* in the plane and to be a planar representation of G. In general, $\tilde{G}$ will denote an embedding of G.

We can extend the idea of embedding to other surfaces. Figure 3.1 (*a*) shows the complete graph with five vertices which, as we shall prove later, cannot be embedded in the plane. Figure 3.1 (*b*) shows that K_5 can in fact be embedded on a toroidal surface. A torus is a solid figure obtained by rotating a circle (or in fact any closed curve) about a line in its plane but not intersecting it.

An inherent property of planar graphs is embodied in the following theorem:

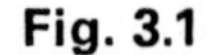

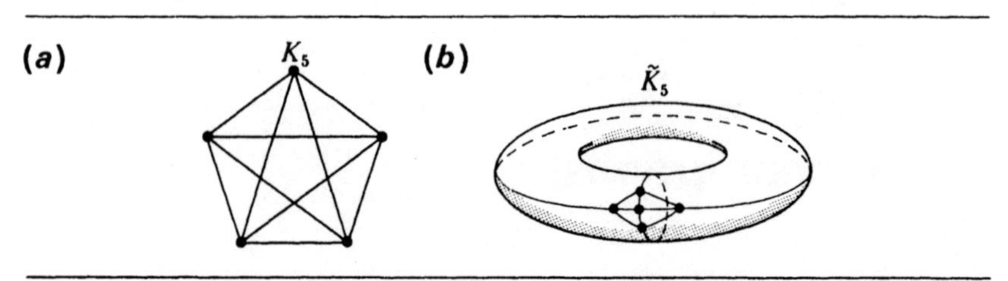

Theorem 3.1. A graph G is embeddable in the plane if and only if it is embeddable on the sphere.

Proof. We show this by using a mapping known as *stereographic projection.* Consider a spherical surface S, touching a plane P at the point x. The point y (called the *point of projection*) is on S and diametrically opposite x. Any point z on P can be projected uniquely onto S at z' by making y, z and z' collinear. In this way any graph embedded in P can be projected onto S. Conversely, we can project any graph embedded in S onto P, choosing y so as not to lie on any vertex or edge of the graph. ■

A planar representation of a graph divides the plane into a number of connected regions, called *faces*, each bounded by edges of the graph. Figure 3.2(*a*) indicates the faces of a particular embedding of the graph shown there. Of course, any planar representation of a (finite) graph always contains one face enclosing the graph. This face, called the exterior *face*, is f_1 in figure 3.2(*a*). Theorem 3.2 will be of particular use later on.

Fig. 3.2

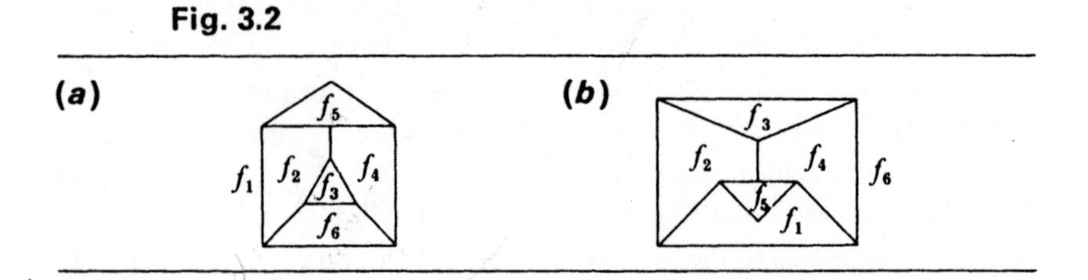

Theorem 3.2. A planar embedding of a graph can be transformed into a different planar embedding such that any specified face becomes the exterior face.

Proof. Any face of $\tilde{G}$ is defined by the path which forms its boundary. Any such path, T, identified in a particular planar representation P of G, may be made to define the exterior face of a different planar representation P' as follows. We form, as we can according to theorem 3.1, a spherical embedding P'' of P. P' is then formed by projecting P'' onto the plane in such a way that the point of projection lies in the face defined by the image of T on the sphere. ■

Figure 3.2(b) shows a mapping of the graph of figure 3.2(a) according to theorem 3.2 so that f_6 becomes the exterior face.

There is a simple formula connecting the number of faces, edges and vertices in a connected planar graph. *Euler's formula*, as it is known, will be of particular use to us in establishing the non-planarity of two important graphs. We derive the formula in theorem 3.3 in which the following notation is used. For a graph $\tilde{G}$, $n(G)$ denotes the number of vertices, $e(G)$ the number of edges and $f(G)$ the number of faces. Where there is no ambiguity we respectively write n, $|E|$ or f.

Theorem 3.3. If G is a connected planar graph, then, for any $\tilde{G}$:

$$f = |E|-n+2$$

Proof. By induction on f. For $f = 1$, G is a tree and by theorem 1.2, $|E| = n-1$, and so the formula holds. Suppose it holds for all planar graphs with less than f faces and suppose that $\tilde{G}$ has $f \geqslant 2$ faces. Let (u, v) be an edge of G which is not a cut-edge. Such an edge must exist because $\tilde{G}$ has more than one face. The removal of (u, v) from $\tilde{G}$ will cause the two faces separated by (u, v) to combine, forming a single face. Hence $(\widetilde{G-(u, v)})$ is a planar embedding of a connected graph with one less face than $\tilde{G}$, hence:

$$f(G-(u, v)) = f(G)-1$$

also

$$n(G-(u, v)) = n(G)$$

and

$$e(G-(u, v)) = e(G)-1$$

But by the induction hypothesis:

$$f(G-(u, v)) = e(G-(u, v))-n(G-(u, v))+2$$

and so, by substitution:

$$f(G) = e(G)-n(G)+2$$

Hence, by induction, Euler's formula holds for all connected planar graphs. ■

We shall require three corollaries to theorem 3.3. Before presenting them, we define the *degree of a face*, $d(f)$, to be the number of edges bounding the face f and we denote the number of vertices of degree i by $n(i)$.

Lemma 3.1. For a simple planar graph G, we have for any $\tilde{G}$:

$$2e(G) = \sum_i d(f_i) = \sum_j jn(j)$$

because each edge contributes one to the degree of each of two vertices.

Corollary 3.1. For any simple connected planar graph G, with $|E| > 2$, the following holds:

$$|E| \leqslant 3n-6$$

Proof. Each face of $\tilde{G}$ is bounded by at least three edges and so:

$$\sum_i d(f_i) \geqslant 3f$$

The result then follows by substitution into Euler's formula and using lemma 3.1. ■

Corollary 3.2. For any simple connected *bipartite* planar graph G, with $|E| > 2$, the following holds:

$$|E| \leqslant 2n-4$$

Proof. Each face of $\tilde{G}$ is bounded by at least four edges. The result then follows as for corollary 3.1. ■

The third corollary will be of particular use in chapter 7.

Corollary 3.3. In a simple connected planar graph there exists at least one vertex of degree at most 5.

Proof. From corollary 3.1:

$$|E| \leqslant 3n-6$$

also $n = \sum_i n(i)$ and from Lemma 3.1, $2|E| = \sum_i i\, n(i)$. Therefore, by substitution:

$$\sum_i (6-i)\, n(i) \geqslant 12$$

The left-hand side of this inequality must clearly be positive. Since i and $n(i)$ are always non-negative it follows that there must exist some non-zero $n(i)$ for at least one i less than six. ■

As examples of the use of corollaries 3.1 and 3.2 we now establish the non-planarity of the two graphs K_5 and $K_{3,3}$. These graphs play a fundamental rôle in one characterisation of planarity, embodied in Kuratowski's theorem which is presented in section 3.3. Now K_5 has five vertices and ten edges and so cannot be planar because the inequality of corollary 3.1 is violated. Similarly, $K_{3,3}$ cannot be planar because with six vertices and nine edges the inequality of corollary 3.2 is not satisfied.

Corollaries 3.1 and 3.2 are necessary but not sufficient to characterise planar graphs and therefore have limited applicability. In section 3.3 we describe ways to more precisely characterise planar graphs.

3.2 Genus, crossing-number and thickness

We have seen that both K_5 and $K_{3,3}$ cannot be embedded in the plane. Both, in fact, are *toroidal* graphs, that is to say that they can be embedded in the surface of a torus. For K_5 this embedding is illustrated in figure 3.1 (*b*). It is instructive to understand the topological difference between a spherical surface and a toroidal surface. Any single closed *line* (or *curve*) embedded in a spherical surface will divide the surface into two regions. On the other hand, a closed curve embedded in a toroidal surface will not necessarily divide it into two regions, although any two non-intersecting closed curves are guaranteed to. Figure 3.3 shows a closed curve C drawn first on a

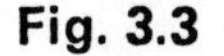
Fig. 3.3

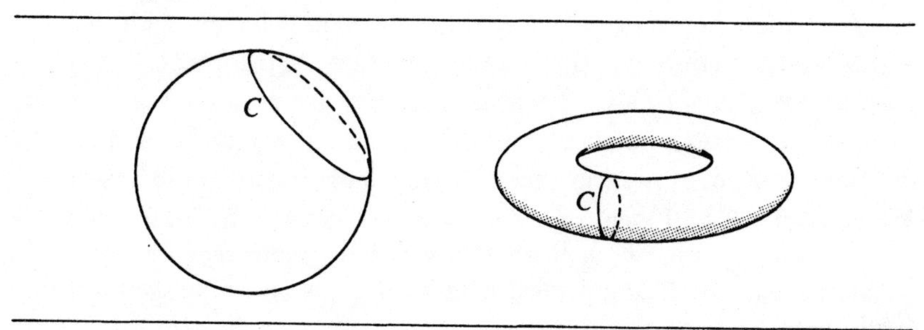

spherical surface and then on a toroidal surface. In the first case two regions result but in the second case the surface remains connected. For any non-negative integer g, we can construct a surface in which it is possible to embed g non-intersecting closed curves without separating the surface into two regions. If for the same surface $(g+1)$ closed curves *always* cause a separation, then the surface is said to have a *genus* equal to g. For a spherical surface $g = 0$, while for a toroidal surface $g = 1$.

The genus is a topological property of a surface and remains the same if the surface is deformed. The toroidal surface is topologically like a spherical surface but with the addition of a 'handle', as shown in figure 3.4. In that diagram $K_{3,3}$ has been embedded on the toroidal surface. Any surface of genus g is topologically equivalent to a spherical surface with g handles. A graph that can be embedded in a surface of genus g, but not on a surface of genus $(g-1)$ is called a graph of genus g. Notice that the so-called *crossing-number* of a graph (that is, the minimum number of crossings of edges for the graph drawn on the plane) is not the same as its

Fig. 3.4

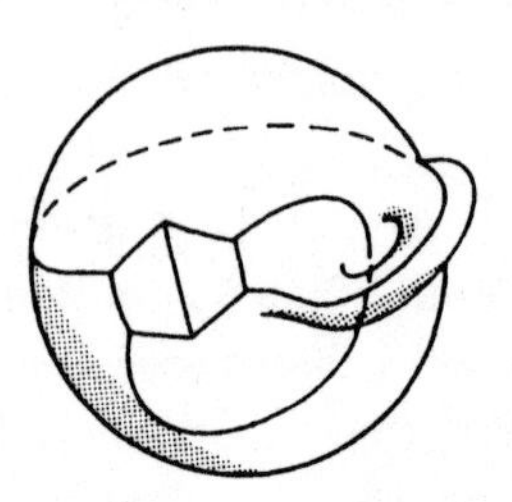

genus. More than one edge may pass over or under a handle on the sphere and so the genus of a graph will not exceed its crossing-number.

Theorem 3.4. If G is a connected graph with genus g, n vertices, $|E|$ edges and if $\tilde{G}$ has f faces, then:

$$f = |E| - n + 2 - 2g$$

Proof. By induction on g. For $g = 0$ the theorem coincides with theorem 3.3. As our induction hypothesis we assume that the theorem is true for all graphs with genus $(g-1)$. These graphs may be drawn on a spherical surface with $(g-1)$ handles and include all those graphs obtained by deleting those edges passing over a single handle in any graph of genus g. We construct G with genus g on a surface of genus g by adding a single edge to some graph G', such that this added edge forces the use of an additional handle. Using primed letters for G', we have by the induction hypothesis:

$$f' = |E'| - n' + 2 - 2g'$$

but

$$|E| = |E'| + 1,\ g = g' + 1 \text{ and } n = n'$$

Also $f = f' - 1$ because the handle connects two distinct faces in G' making a single face in G. Hence by substitution:

$$f = |E| - n + 2 - 2g$$

and so by induction the theorem is proved. Notice that adding more (non-crossing) edges over the handle does not change the genus of the graph, although each edge added in this way also adds another face to the graph so that the formula continues to hold true. ■

Genus and crossing-number have obvious implications for the manufacture of electrical circuits on planar sheets. A fact of recent interest for the large scale integrated circuits of silicon chips is that there is (Lipton & Tarjan[11]) a planar equivalent for any *boolean* electrical circuit, obtained by replacing each pair of crossing wires by a fixed size planar subcircuit which simulates the behaviour of the original crossing. Figure 3.5 shows such a simulation using, in this case, **exclusive- or** gates. The cross-over of wires (X, X') and (Y, Y') of (*a*) is replaced by three **exclusive- or** gates in (*b*). It is easy to check that whatever boolean values are input at X and Y they will be reproduced respectively at X' and Y'. In a planar circuit with straight wire connections and n vertices (gates), there can be at most $O(n^2)$ cross-overs. Hence a planar equivalent of a boolean circuit can be obtained at the expense of at most an $O(n^2)$ increase in the number of gates (the so-called circuit size).

Fig. 3.5

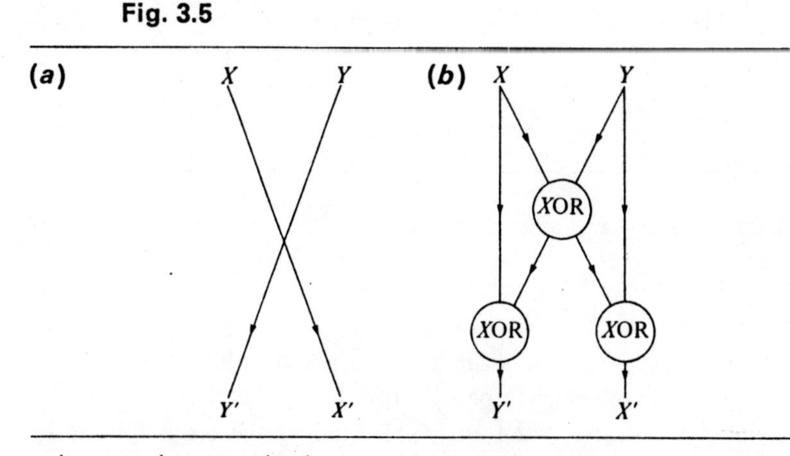

A convenient practice is to make connections between parallel planar subcircuits separated by insulating sheets at vertices of the corresponding graph. The problem is then equivalent to decomposing the graph into planar subgraphs and, in particular, we become interested in the so-called *thickness* of a graph. The thickness $T(G)$ of a graph G is the minimum number of planar subgraphs of G whose *union* is G. If $G_1 = (V, E_1)$ and $G_2 = (V, E_2)$, then their union, $G_1 \cup G_2$, is the graph $(V, E_1 \cup E_2)$. Figure 3.6 shows three graphs G_1, G_2 and G_3 whose union is K_9. Hence $T(K_9) \leqslant 3$. We shortly present an expression for the thickness of a complete graph on n vertices and this provides an upper bound for any graph with the same number of vertices.

Before completing this section we note two corollaries arising from theorem 3.3 and from theorem 3.4.

Corollary 3.4. The thickness T of a simple graph with n vertices and $|E|$ edges satisfies:

$$T \geqslant \left\lceil \frac{|E|}{3n-6} \right\rceil$$

Proof. Each planar subgraph will contain, according to corollary 3.1, at most, $(3n-6)$ edges and so the result follows. ■

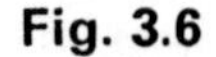

Fig. 3.6

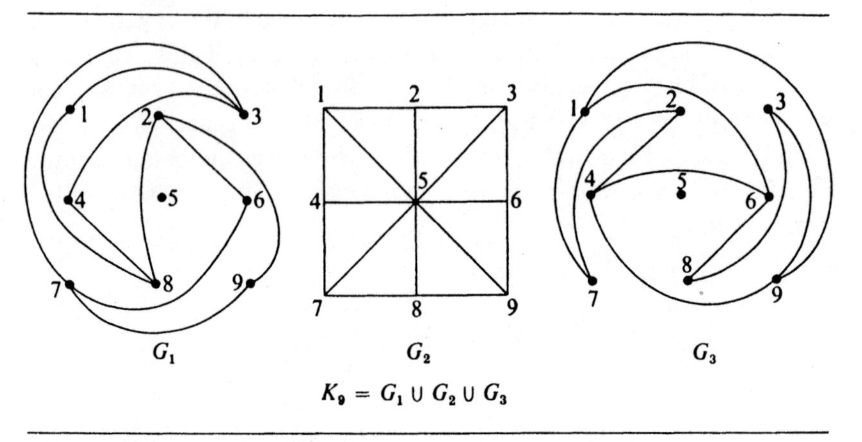

$K_9 = G_1 \cup G_2 \cup G_3$

Corollary 3.5. The genus g of a simple graph with n ($\geqslant 4$) vertices and $|E|$ edges satisfies:

$$g \geqslant \lceil \tfrac{1}{6}(|E| - 3n) + 1 \rceil$$

Proof. Every face of an embedding of the graph is bound by at least three edges each of which separates two faces, therefore $3f \leqslant 2.|E|$. From theorem 3.4, $g = \frac{1}{2}(|E| - n - f) + 1)$ and so the result follows by substitution. ■

Specific results for thickness and genus are known for special cases (e.g., complete graphs, complete bipartite graphs (see exercise 3.11)) and involve lengthy proofs. In the case of complete graphs $|E| = \frac{1}{2}n(n-1)$ and the above corollaries then give:

$$g \geqslant \lceil \tfrac{1}{12}(n-3)(n-4) \rceil$$

and

$$T \geqslant \left\lceil \frac{n(n-1)}{6(n-2)} \right\rceil = \left\lfloor \frac{n(n-1)+(6n-14)}{6(n-2)} \right\rfloor = \lfloor \tfrac{1}{6}(n+7) \rfloor$$

It is known that in the result for g equality holds. Similarly, equality holds in the expression for T except for $n = 9$ and for $n = 10$, in both cases $T = 3$. These refinements required the considerable efforts of mathematicians over many years. Beineke & Wilson[7] provides a reference list of primary sources.

Filotti *et al.*[18] have described an $O(n^{O(g)})$-algorithm which takes as input a graph G and a positive integer g and which then finds an embedding of G on a surface of genus g if such an embedding exists.

3.3 Characterisations of planarity

In section 3.1 we proved that K_5 and $K_{3,3}$ are non-planar. These two graphs play a fundamental rôle in the classical characterisation of planarity due to Kuratowski and which is embodied in theorem 3.5. We use Kuratowski's theorem to establish two other descriptions of planarity which more precisely fit the requirements of this text. Before proceeding we need some definitions.

By $G_1 = (V_1, E_1)$ we denote a subgraph of $G = (V, E)$. A *piece* of G relative to G_1 is then:

either

(*a*) an edge $(u, v) \in E$ where $(u, v) \notin E_1$ and $u, v \in V_1$,

or

(*b*) a connected component of $(G - G_1)$ plus any edges incident with this component.

In figure 3.7 the graph G has a subgraph G_1 which is a circuit $(v_1, v_2, v_3, v_4, v_5, v_1)$. B_1, B_2 and B_3 are the pieces of G relative to G_1. For any piece B, the vertices which B has in common with G_1 are called the *points of contact* of B. Thus in figure 3.7 B_1 has the points of contact v_3 and v_5, while B_3 has the points of contact v_1, v_2 and v_5. If a piece has two or more points of contact then it is called a *bridge*. Thus B_1 and B_3 are bridges but B_2 is not a bridge.

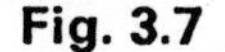
Fig. 3.7

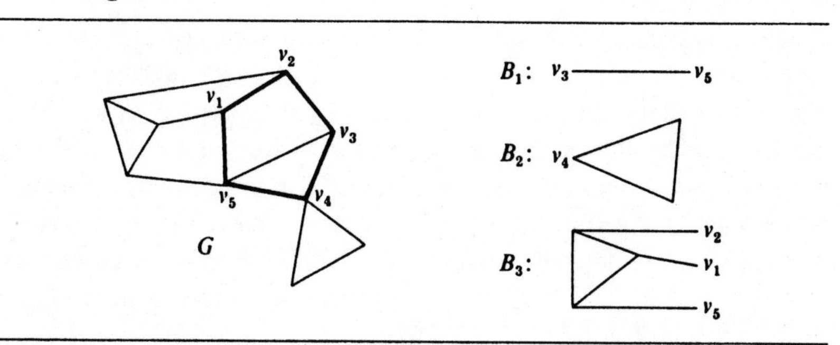

Obviously a graph is planar if and only if each of its blocks is planar. Thus in questions of planarity we can always assume that we are dealing with blocks. Any piece of a block with respect to any proper subgraph is clearly a bridge.

Let C be any circuit which is a subgraph of G. $\tilde{C}$ then divides the plane into two faces, an *interior* face and an *exterior* face. For every pair of vertices of a given bridge of C, there is a path from one vertex to the other which does not use an edge of C. Of course, if G is planar, and if there exists a single bridge relative to C, then C is a boundary of some face because the bridge can belong to one and only one (namely, the other) face of C. Two bridges B_1 and B_2 are said to be *incompatible* ($B_1 \not\approx B_2$) if, when placed in the same face of the plane defined by C, at least two of their edges cross. See figure 3.8(*a*). To establish incompatibilities, each bridge is conveniently reduced to a single vertex connected to the points of contact with C.

Fig. 3.8

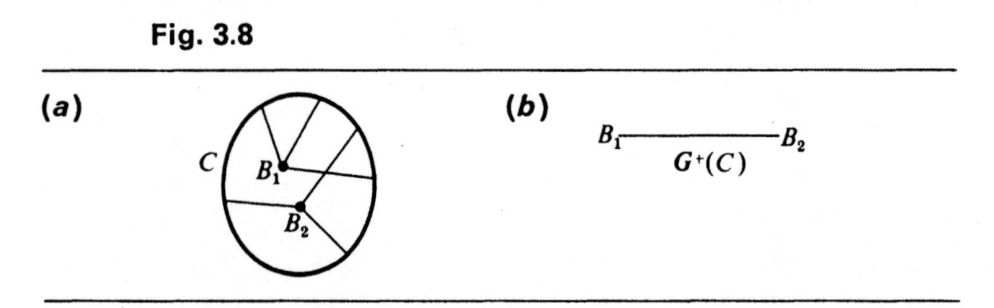

An *auxiliary graph* $G^+(C)$ relative to a circuit C has a vertex-set consisting of a vertex for each bridge relative to C and an edge between any two such vertices B_i and B_j if and only if $B_i \not\approx B_j$. See, for example, figure 3.8(*b*). Suppose that $G^+(C)$ is a bipartite graph with bipartition $(B, \overline{B})$. Then the bridges in B may be embedded in one face of C and the bridges in $\overline{B}$ may be embedded in the other face. In this way no incompatible bridges occur in the same face.

Before presenting Kuratowski's theorem we need just one more definition. Whether or not a graph is planar is obviously unaffected either by dividing an edge into two edges in series by the insertion of a vertex of degree 2, or by the reverse of this process. Two graphs are said to be *homeomorphic* if one can be made isomorphic to the other by the addition or the deletion of vertices of degree two in this manner. Figure 3.9(*a*) shows a graph which is homeomorphic to $K_{3,3}$, while (*b*) shows a graph which

Fig. 3.9

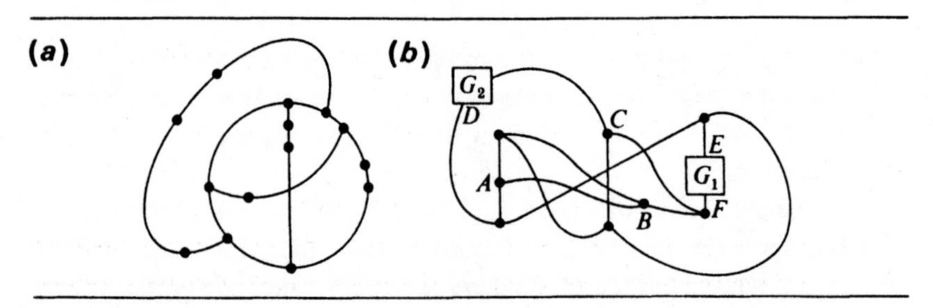

contains a subgraph homeomorphic to $K_{3,3}$. In this second case the subgraph is obtained by deleting the edge (A, B), by replacing the connected subgraph G_1 by the path it contains from E to F and by similarly replacing the connected subgraph G_2 by a path from D to C.

Theorem [Kuratowski] 3.5. A graph is planar if and only if it has no subgraph homeomorphic to K_5 or to $K_{3,3}$.

Proof. In section 3.1 we proved that K_5 and $K_{3,3}$ are non-planar. It follows that any graph containing a subgraph homeomorphic to either cannot be planar.

It remains to be shown that a graph is planar if it does not contain a subgraph homeomorphic to K_5 or to $K_{3,3}$. We shall prove this by induction on the number of edges. It is clearly true for graphs with one or two edges. As the induction hypothesis we assume it to be true for all graphs with less than N edges. We now show that it is true for the graph G with N edges by demonstrating that the following statement leads to a contradiction: G is non-planar and does not contain a subgraph homeomorphic to K_5 or to $K_{3,3}$.

If G is non-planar, the following consequences apply:

(*a*) G must be connected. Otherwise G would consist of a number of components each with less than N edges, and each not having a subgraph homeomorphic to K_5 or $K_{3,3}$ (because G does not). By the induction hypothesis each component would be planar and hence so would G.

(*b*) G must not contain a point of articulation. If it did then G could be separated at this point of articulation, x. Each resulting component would be planar as in (*a*). For each component x could be mapped into the exterior face of a planar embedding according to theorem 3.2. The components could then clearly be rejoined at x without loss of planarity. Hence G would be planar.

(*c*) If any edge of G is removed, say (x, y), then the remaining graph G' contains a simple circuit passing through x and y. Notice that G' is connected because G contains no point of articulation. If no such simple circuit exists then every path from x to y would have to pass through a common vertex, say z. In other words, z would be an articulation point of G'. G' could then be separated at z into two components, G_1' (containing x) and G_2' (containing y). We add the edge (x, z) to G_1' so forming G_1'', and we add the edge (y, z) to G_2', so forming G_2''. Now neither G_1'' nor G_2'' could contain subgraphs homeomorphic to K_5 or to $K_{3,3}$ otherwise G would. This is because G contains a subgraph homeomorphic to G_1'', for example, where the path $(x, y, ..., z)$ in G takes the part of (x, z) in G_1''. By the induction hypothesis G_1'' and G_2'' would be planar. According to theorem 3.2 we could map (x, z) of G_1'' into the boundary of the exterior face of $\tilde{G}_1''$, similarly, we could take (y, z) of G_2'' to the exterior face of $\tilde{G}_2''$. Without loss of planarity, the two graphs G_1'' and G_2'' could then be joined at z and the edges (x, z) and (y, z) replaced by (x, y). This planar reconstruction of G thus yields a contradiction and so G' cannot contain an articulation point. G' is thus a block and so by theorem 2.10 contains a simple circuit passing through x and y.

Thus, summarising, $G' = G-(x, y)$ is connected and contains a simple circuit C passing through x and y. In fact C could be one of a number of such circuits. G' contains no subgraph homeomorphic to K_5 or to $K_{3,3}$, has one less edge than G and so, by the induction hypothesis, is planar. Let $\tilde{G}'$ be a planar embedding of G'. We then choose C to be the circuit passing through x and y which contains the largest number of faces of $\tilde{G}'$ in its interior. Any bridge of G' with respect to C is called an *interior* or an *exterior* bridge depending upon whether it lies in the interior or exterior of C for the embedding $\tilde{G}'$. For convenience we assign a direction to C which we take to be clockwise. If p and q are vertices on C, then $S[p, q]$ denotes the set of vertices from p to q (including p and q) on S going in a clockwise direction. $S]p, q[$ denotes $S[p, q]-\{p, q\}$. Note that no exterior bridge can have more than one point of contact in $S[x, y]$ or in $S[y, z]$. Otherwise C could be expanded to enclose at least one more face of $\tilde{G}'$.

G is constructed from the planar graph G' by adding the edge (x, y). Consider the requirements of exterior and interior bridges of $\tilde{G}'$ with respect to C in order that G be non-planar. There must exist at least one exterior bridge E and one interior bridge I. As fas as E is concerned there will be just two points of contact i and j with C such that:

$$i \in S]x, y[\quad \text{and} \quad j \in S]y, x[$$

I may have any number of points of contact with C. We certainly require that there are points of contact:

$$a \in S]x, y[\quad \text{and} \quad b \in S]y, x[$$

otherwise (x, y) may be added to the interior of C. We also require points of contact:

$$c \in S]i, j[\quad \text{and} \quad d \in S]j, i[$$

in order that $I \not\approx E$. In other words, I must be incompatible with E so that it cannot be taken into the exterior of C without loss of planarity.

Figure 3.10 schematically illustrates this. In this diagram a coincides with c and b coincides with d. There are however other possible configurations. Figure 3.11 illustrates all of those that are essentially different. For reasons of

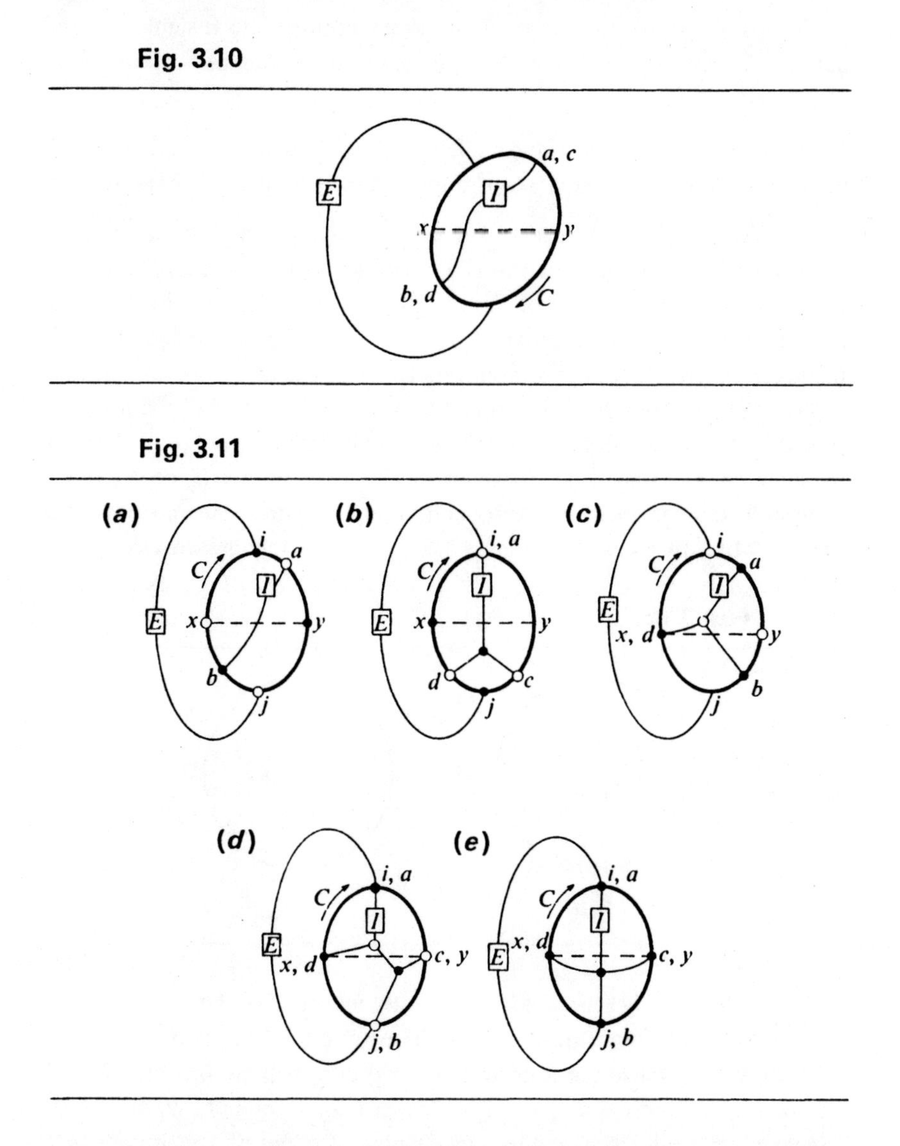

Fig. 3.10

Fig. 3.11

clarity whenever any of a, b, c or d coincide, a single label is used. Notice that the configurations (d) and (e) differ only according to the internal paths in I linking a, b, c and d. Each of the configurations illustrated in (a), (b), (c) and (d) exhibit subgraphs which are homeomorphic to $K_{3,3}$. Open and closed circles are used to indicate the vertices of each partition.

The rather exceptional case indicated in (*e*) exhibits a subgraph homeomorphic to K_5. We have thus found the contradiction we were seeking and so the theorem is proved. ■

The following theorem provides a more appropriate insight into the nature of planarity as far as the planarity algorithm of section 3.4 is concerned.

Theorem 3.6. A necessary and sufficient condition for a graph G to be planar is that for every circuit C of G the auxiliary graph $G^+(C)$ is bipartite.

Proof. The condition is necessary because for any circuit C of a planar graph G, we can form a bipartition $(B, \bar{B})$ of the bridge vertices of G relative to C, such that bridges in B lie in one face of C for $\tilde{G}$, and the bridges of $\bar{B}$ lie in the other face. Clearly, $G^+(C)$ is bipartite because no edge of $G^+(C)$ connects two vertices in B or connects two vertices in $\bar{B}$.

That the condition is sufficient can be seen as follows. If G is not planar then according to Kuratowski's theorem G contains a subgraph homeomorphic to K_5 or to $K_{3,3}$. We suppose that G contains K_5 or $K_{3,3}$ as a subgraph, the generalisation to G containing proper homeomorphisms is obvious. In either case (see figure 3.12, in which the chosen circuits are

Fig. 3.12

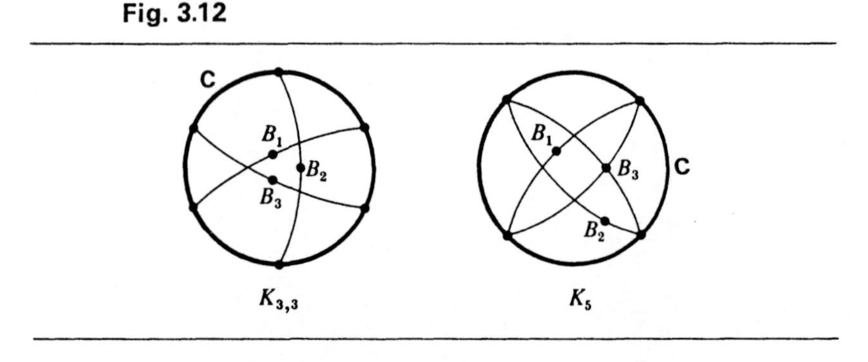

indicated by heavily scored edges), we can choose C of the subgraph such that $G^+(C)$ is not bipartite. For $K_{3,3}$ there are three bridges B_1, B_2 and B_3, each of which is a single edge and any two of which are incompatible. In the case of K_5 there are again three bridges B_1, B_2 and B_3. B_1 and B_2 are single edges while B_3 is a vertex of K_5 plus its edges of attachment to C. Again any two of the bridges are incompatible. Thus for both K_5 and $K_{3,3}$, for the circuits chosen, $G^+(C) = K_3$ which is not bipartite. ■

The second characterisation of planarity of particular use in this text concerns *dual* graphs to which we devote the following section.

3.3.1. Dual graphs

The main purpose of this section is to provide an alternative way to characterise planar graphs. In particular we shall see that a graph is planar if and only if it has a *dual*. However, as we shall see, there is an important connection between circuits of a graph and the cut-sets of its dual. This connection provides an additional stimulus for our interest in dual graphs.

Given a particular planar representation $\tilde{G}$ of a graph, we informally introduce the idea of its dual G^* by providing construction rules for it. A vertex of G^* is associated with each face of $\tilde{G}$. For each edge e_i of $\tilde{G}$ there is an associated edge e_i^* of G^*. If e_i separates the faces f_j and f_k in $\tilde{G}$, then e_i^* connects the two vertices of G^* associated with f_j and f_k. Exceptionally, e_i may not separate two faces of $\tilde{G}$, namely, when e_i is incident with a vertex of degree one. In this case e_i^* forms a self-loop on the vertex of G^* associated with the face of $\tilde{G}$ surrounding e_i. An example construction is shown in figure 3.13. We do specify which of the two overlain graphs is the dual. In fact either one is the dual of the other as can be easily verified by inspection. This is a consequence of the construction process and not of the example. Notice that that G^* must also be planar.

Fig. 3.13

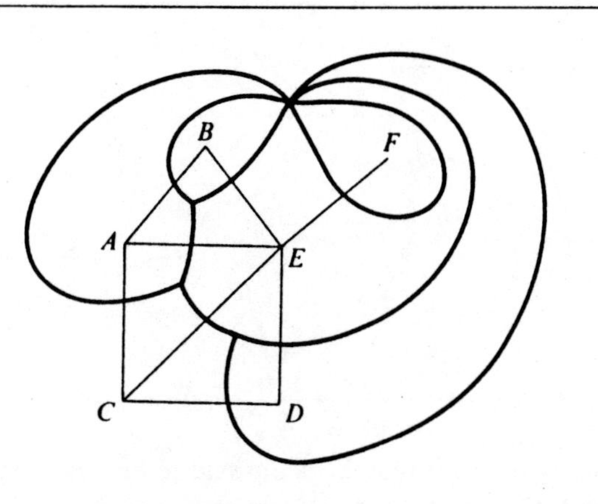

We have carefully referred to the dual of a *planar representation* $\tilde{G}$ of a graph G and *not* to the dual of that graph. Figure 3.14 contains a different planar representation of a graph first shown in figure 3.13. We can see that the dual of the second representation is not isomorphic to the dual of the first representation. In particular the vertex X in figure 3.14 has degree six unlike any vertex in figure 3.13. In fact, there is a rather simple constructional relationship between the duals of different planar representations

Fig. 3.14

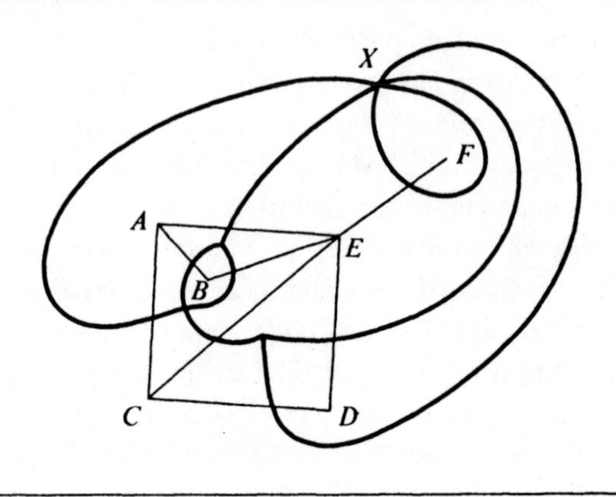

of the same graph. We illustrate this in figure 3.15, where (*a*) illustrates a graph isomorphic to the unlabelled graph in figure 3.13. In (*b*) this has been separated into two components by division of the vertices A and B. Figure 3.15(*c*) shows a graph isomorphic to the graph containing the vertex X in figure 3.14. This has been constructed by identifying vertex A_1 with B_2 and vertex A_2 with B_1 in figure 3.15(*b*).

Fig. 3.15

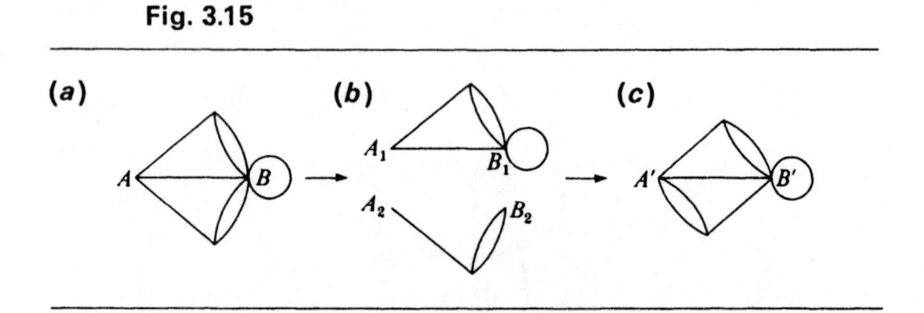

The graphs, figure 3.15(*a*) and (*c*), are said to be *2-isomorphic*. Any two graphs G_1 and G_2 are 2-isomorphic if they become isomorphic under repeated application of either or both of the following operations:

(*a*) separation of G_1 or G_2 into two or more components at articulation points,

(*b*) if G_1 and G_2 can be divided into two disjoint subgraphs with two vertices in common, then separate at these vertices, A and B, and reconnect so that A_1 coincides with B_2 and A_2 coincides with B_1 as in figure 3.15.

As a further example, the two graphs of figure 3.16 are 2-isomorphic.

Fig. 3.16

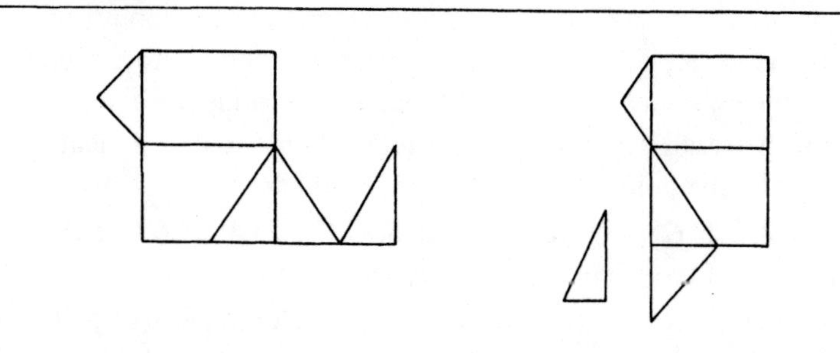

We state the following theorem without proof (see excercise 3.10).

Theorem 3.7. All the duals of a planar graph G are 2-isomorphic and any graph 2-isomorphic to a dual of G is also a dual of G.

We require a (combinatorial) definition of a dual graph which will suit our purposes in a better way than the (geometric) definition outlined earlier. This is provided as follows:

Definition of a dual of a graph. Let G_1 and G_2 denote graphs with a one-to-one correspondence between their edges and let C denote the set of edges forming *any* simple circuit in G_1. G_2 is a *dual* of G_1 if and only if the corresponding set of edges C^* in G_2 is a cut-set.

Notice that this definition makes no allusion to G_1 or G_2 being planar. We shall however prove that if G_1 is planar then the above combinatorial definition coincides with the geometric definition:

Theorem 3.8. Every planar graph has a (planar, combinatorial) dual.

Proof. It is clear that every planar graph has a (planar, geometric) dual. Given a planar graph we construct its geometric dual G^* overlaying $\tilde{G}$ in the manner described earlier. Any simple circuit C of $\tilde{G}$ divides the plane into two regions and so the vertices of G^* are divided into two (non-empty) subsets. Removal of the set of edges C^* of G^* (which cross C in $\tilde{G}$) clearly separates G^* into two components. Hence C^* is a cut-set of G^*.

Similarly any cut-set C^* of G^* defines a corresponding set of edges in C in $\tilde{G}$. We shall show that if C is not a simple circuit then C^* cannot be a cut-set. By construction only one vertex of G^* sits in each face of $\tilde{G}$. Consider the set of edges radiating from a single vertex of $\tilde{G}$. Each such edge separates two faces of $\tilde{G}$ each containing a vertex of G^*. The end-points of the corresponding edges of G^* have no choice of which vertex to be attached to, and these edges are thus constrained to form the boundary of a face of G^*. Hence every vertex of $\tilde{G}$ sits in one face of G^*. If C is not a circuit in $\tilde{G}$ then there are at least two edges of C with end-points not connected to others in C. If C^* is a cut-set then these end-points of C must lie in the same face of G^*. But this is a contradiction because each face of G^* contains only one vertex of $\tilde{G}$. Notice that C must be a *simple* circuit otherwise it would separate G^* into more than two components and so C^* would be the union of more than one cut-set. ■

Corollary 3.6. If G^* is a dual of $\tilde{G}$ then $\tilde{G}$ is a dual of G^*. The proof is straightforward and similar to that for theorem 3.8.

From now on, when we refer to the dual of a graph, we shall have in mind the definition of a combinatorial dual. We remember that this definition makes no reference to planarity. The following theorem is the main result of this section.

Theorem 3.9. A graph has a dual if and only if it is planar.

Proof. From theorem 3.8 we know that every planar graph has a dual. We need, therefore, only to prove that a non-planar graph has no dual. From the definition of dual it is clear that a graph G can only have a dual if every subgraph of G has a dual. Also if a graph has a dual then any graph homeomorphic to it must have a dual. Since every non-planar graph contains, according to theorem 3.5, a subgraph homeomorphic to K_5 and/or to $K_{3,3}$ we need only show that these graphs have no dual. We do this in (i) and (ii) below:

(i) We suppose that K_5 has a dual, K_5^*, and show that this leads to a contradiction. We observe that K_5 has ten edges, no circuit of length 2, no cut-set with two edges and cut-sets with only four and six edges. These, respectively, have the following consequences. K_5^* has ten edges, no vertex with degree less than 3, no circuit of length 2 and circuits of length 4 and 6 only. It is easy to see that these are mutually incompatible and so we have the desired contradiction.

(ii) We now suppose that $K_{3,3}$ has a dual, $K_{3,3}^*$ and shall similarly show that this leads to a contradiction. $K_{3,3}$ has no cut-set consisting of two edges and so $K_{3,3}^*$ has no circuits of length 2. Also $K_{3,3}$ has circuits of length 4 and 6 only, therefore $K_{3,3}^*$ has no cut-set with less than four edges. It follows that the degree of every vertex in $K_{3,3}^*$ is at least 4. That is, there are at least five vertices in $K_{3,3}^*$, each of at least degree 4, requiring $\frac{1}{2}(5 \times 4) = 10$ edges. However, $K_{3,3}^*$ must have the same number of edges as $K_{3,3}$, that is, nine. Thus we have found the required contradition. ■

We conclude this section by anticipating an interest in dual graphs that arises in chapter 7. In that chapter we turn our attention to the problem of colouring areas of a map using the minimum number of colours such that no two adjacent regions are similarly coloured. It is now known that the famous 'four-colour' conjecture is true, namely, that four colours are sufficient. All we wish to note here is that the map colouring problem is precisely equivalent to the problem of colouring the vertices of the dual (of the graph corresponding to the map) such that no two adjacent vertices are similarly coloured. The dual graph provides a more convenient vehicle for reasoning about the problem.

3.4 A planarity testing algorithm

Before subjecting a particular graph to an algorithm which determines whether or not it is planar, some preprocessing may considerably simplify the task. In this connection we note the following points:

(*a*) If the graph is not connected then we subject each component to the test separately.
(*b*) If the graph is separable (that is, has one or more articulation points) then it is clearly planar if and only if each of its blocks is planar. We therefore disconnect the graph and subject each block separately to the test.
(*c*) Self-loops may obviously be removed without affecting planarity.
(*d*) Each vertex of degree 2 plus its incident edges can be replaced by a single edge. In other words, we construct the homeomorphic graph with the smallest number of vertices. This graph is clearly planar if and only if the original graph is planar.
(*e*) Parallel edges can clearly be removed without affecting planarity.

The last two simplifying steps ought to be applied repeatedly and alternately until neither can be applied further. Following these simplifications two elementary tests can be applied:

(*f*) If $|E| < 9$ or $n < 5$ then the graph must be planar.
(*g*) If $|E| > 3n-6$ then the graph, by corollary 3.1, must be non-planar.

If these two tests fail to resolve the question of planarity then the pre-processed graph is subjected to a more elaborate test. We pursue that shortly. First it is worth demonstrating what simplification can result from this preprocessing, particularly the repeated applications of (*d*) and (*e*). Figure 3.17 shows a graph with three blocks subjected to this processing which resolves that the graph is planar.

Many algorithms have been published which test for planarity. Planarity testing can be done in O(n) time as Hopcroft & Tarjan[2] first showed.

Fig. 3.17

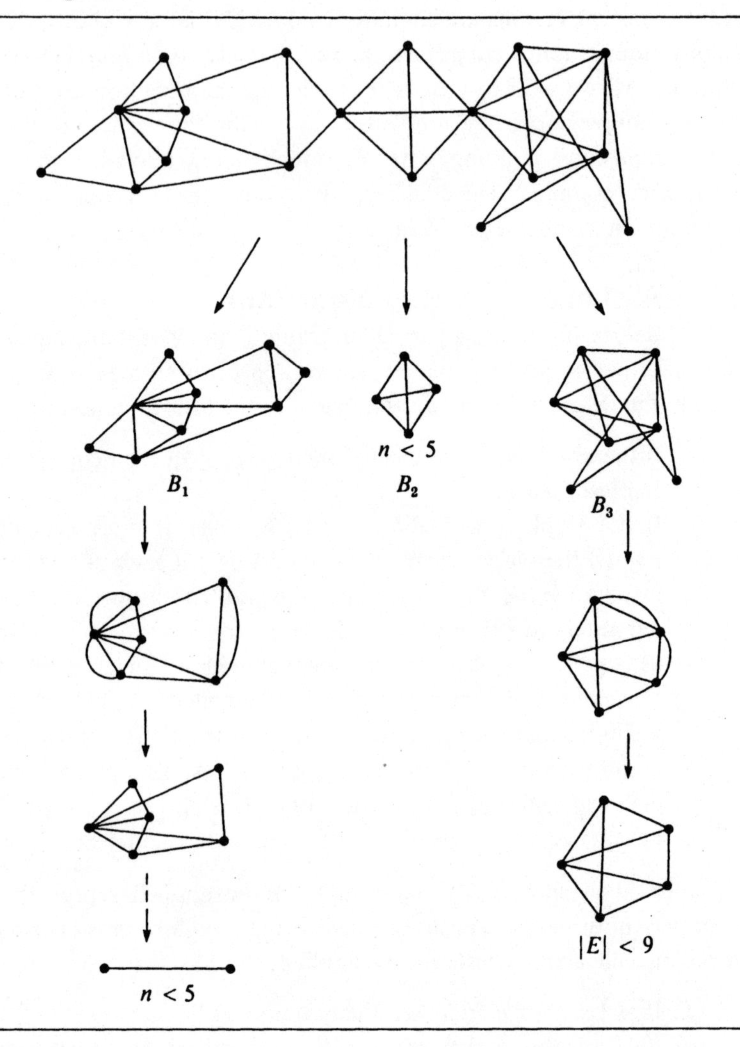

Lempel, Even & Cederbaum[3] published an algorithm which, through the work of Even & Tarjan[4] and Leuker & Booth[5] was also shown to be realisable in O(n)-time. These two algorithms require lengthy explanations and verification. We therefore describe a much simpler but nevertheless fairly efficient algorithm due to Demoucron, Malgrange & Pertuiset.[6] Of course, what is subjected to the algorithm, following any preprocessing, is a block. Before describing the algorithm we need one further definition.

Let $\tilde{H}$ be a planar embedding of the subgraph H of G. If there exists a planar embedding $\tilde{G}$, such that $\tilde{H} \subseteq \tilde{G}$, then $\tilde{H}$ is said to be *G-admissible.*

For example consider figure 3.18. In (*a*) a graph is shown while (*b*) and (*c*) show two different planar embeddings of the same subgraph $H = G-(1, 5)$. In (*b*) $\tilde{H}$ is G-admissible whilst (*c*) shows an embedding of H which is *not* G-admissible.

Fig. 3.18

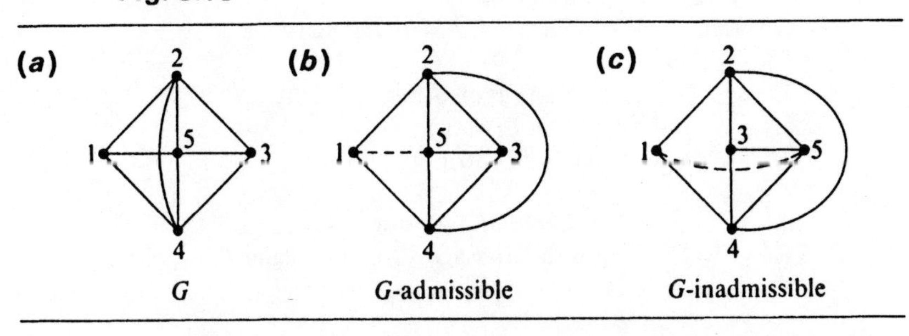

Let B be any bridge of G relative to H. Now, B can be *drawn* in a face of $\tilde{H}$ if all the points of contact of B are in the boundary of f. By $F(B, H)$ we denote the set of faces of $\tilde{H}$ in which B is drawable.

The planarity testing algorithm is outlined in figure 3.19. The algorithm finds a sequence of graphs $G_1, G_2, \ldots$, such that $G_i \subset G_{i+1}$ and finds their planar embeddings $\tilde{G}_1, \tilde{G}_2, \ldots$. If G is planar then, as we shall see, each $\tilde{G}_i$ found by the algorithm is G-admissible and the algorithm terminates with a planar embedding of G, $\tilde{G}_{|E|-n+1}$. If G is non-planar then the algorithm stops with the discovery of some bridge B (with respect to the current G_i) for which $F(B, \tilde{G}_i) = \varnothing$. Obviously a necessary condition that $\tilde{G}_i$ is G-admissible is that for every bridge B relative to G_i, $F(B, \tilde{G}_i) \neq \varnothing$.

The first of the sequence of graphs found by the algorithm, G_1, is a circuit (lines 1–3). Since G is a block it must contain such a circuit. Clearly, G_1 will be planar. The boolean variable *EMBEDDABLE* (lines 5, 6, 10 and 12) has the value **true** so long as the algorithm has not detected a bridge B relative to the current $\tilde{G}_i$ for which $F(B, \tilde{G}_i) = \varnothing$. If it acquires the value **false** then the algorithm terminates (line 6) with the message 'G is non-planar' (line 11). The variable f is used to record the number of faces of the current $\tilde{G}_i$. It is initialised to the value 2 in line 4 and is incremented by one for each execution of the **while** body (lines 7–19). Each execution of the **while** body constructs a new $\tilde{G}_{i+1}$ from the current $\tilde{G}_i$. This is achieved as follows. Lines 7 and 8, respectively, find the set of bridges of G relative to G_i and for each such bridge B, the set $F(B, \tilde{G}_i)$. If there now exists a bridge B which can be drawn in *only one* face F of $\tilde{G}_i$ (i.e., $|F(B, G_i)| = 1$, line 13), then $\tilde{G}_{i+1}$ is constructed by drawing a path P_i between two points of contact of B in the face F. If no such bridge exists

Fig. 3.19. A planarity testing algorithm.

```
1.  Find a circuit C of G
2.  i ← 1
3.  G_1 ← C, G̃_1 ← C
4.  f ← 2
5.  EMBEDDABLE ← true
6.  while f ≠ |E| − n + 2 and EMBEDDABLE do
        begin
7.      find each bridge B of G relative to G_i
8.      for each B find F(B, G̃_i)
9.      if for some B, F(B, G̃_i) = ∅ then
            begin
10.         EMBEDDABLE ← false
11.         output the message 'G is non-planar'
            end
12.     if EMBEDDABLE then
            begin
13.         if for some B, |F(B, G̃_i)| = 1 then F ← F(B, G̃_i)
                else let B be any bridge and F be any face such
                    that F ∈ F(B, G̃_i)
14.         find a path P_i ⊆ B connecting two points of contact
                of B to G_i
15.         G_{i+1} ← G_i + P_i
16.         Obtain a planar embedding G̃_{i+1} of G_{i+1} by drawing P_i
                in the face F of G̃_i
17.         i ← i + 1
18.         f ← f + 1
19.         if f = |E| − n + 2 then output the message 'G is planar'
            end
        end.
```

then P_i is a path between two points of contact for *any* bridge. In either case, P_i divides some face F into two faces and f is incremented by one (line 18). Notice that if G is planar then $\tilde{G}$ will have, according to theorem 3.3, $(|E|-n+2)$ faces and this fact is used to terminate the algorithm (lines 6 and 19). In a more detailed encoding of the algorithm, each $\tilde{G}_i$ may be represented by its set of faces $\{F_i\}$. Here each F_i can be described by the ordered set of vertices which mark its boundary in, say, a clockwise direction about an axis passing through the face. In this sense of course, each axis ought to be viewed from the same side of the plane.

Of course, if the graph is planar, then the algorithm obtains a planar embedding, $G_{|E|-n+1}$, and this could be output in the form of a set of faces by a modification of the conditional statement 19.

Theorem 3.10. The algorithm of Demoucron *et al.* is valid.

Proof. We have to show that each term of the sequence $\tilde{G}_1, \tilde{G}_2, \ldots, \tilde{G}_{|E|-n+1}$, if G is planar, is G-admissible. The proof is by induction. If G is planar then $\tilde{G}_1$ is clearly G-admissible. We assume that $\tilde{G}_i$ is G-admissible for $1 \leqslant i \leqslant k < |E|-n+1$. We now show that $\tilde{G}_{k+1}$ will be G-admissible. Let B and F be as defined in statement 13 of the algorithm. Let $\tilde{G}$ be a planar embedding of G where $\tilde{G}_k \subset \tilde{G}$. If $|F(B, \tilde{G}_k)| = 1$ then, clearly, $\tilde{G}_{k+1}$, as constructed by the algorithm satisfies $\tilde{G}_{k+1} \subseteq \tilde{G}$. We therefore suppose that $|F(B, \tilde{G}_k) > 1$ and imagine that B is *not* drawn in F in $\tilde{G}$ but in some other face F'. Now G is a block so that every bridge of G with respect to G_k has at least two points of contact and can therefore be drawn in just two faces. Thus each bridge with points of contact on the boundary between the faces F and F' may be drawn individually in either F or in F'. Now there clearly exists another planar embedding of G in which each such bridge is drawn in F if it appears in F' in $\tilde{G}$ and is drawn in F' if it appears in F in $\tilde{G}$. The $\tilde{G}_{k+1}$ constructed by the algorithm is clearly G-admissible, since $\tilde{G}_{k+1}$ is contained in this new $\tilde{G}$. ∎

It is easy to see that the planarity testing algorithm can be implemented in polynomial time although it is less sophisticated than the linear-time algorithms mentioned earlier. We leave the details to the reader (exercise 3.14). However we note the following. The body of the *while* statement (lines 7–19) is executed at most $(|E|-n+1)$ times. In order to find each bridge B of $G = (V, E)$ relative to $G_i = (V_i, E_i)$ in line 7, we define $G' = (G-V_i)$, and then need to find:

(*a*) each $(u, v) \in E$ such that $(u, v) \notin E_i$, but $u \in V_i$ and $v \in V_i$,

and

(*b*) each component of G' and add to each component any edges that connect it to vertices in V_i.

For each bridge we need to record its points of contact with G_i. If b is the set of points of contact of B, then in line 8, a face F is in $F(B, \tilde{G}_i)$ if and only if *every* element of b is in F. Here we presume that F denotes an (ordered) set of vertices as described earlier. If each face is described in this manner, then in line 16 $\tilde{G}_{i+1}$ is easily obtained from $\tilde{G}_i$ by simply replacing one $F \in \tilde{G}_{i+1}$ by two new faces in an obvious manner. Returning to the determination of bridges in line 7, notice that all but one of the bridges relative to G_i are bridges relative to $\tilde{G}_{i+1}$. This exceptional bridge is replaced by none or more other bridges. All other steps of the algorithm are easily implemented in an efficient manner.

Figure 3.20 shows an application of the algorithm to the graph G shown there. For each successive G_i, the diagram contains a tabulation of the set of bridges relative to G, the value of f, $F(B, \tilde{G}_i)$, B and F as defined in

Fig. 3.20. An application of the planarity testing algorithm.

$\tilde{G}_i$	f	Bridges	$F(B, \tilde{G}_i)$	B	F	P_i
$\tilde{G}_1$	2	B_1	$\{F_1, F_2\}$			
		B_2	$\{F_1, F_2\}$			
		B_3	$\{F_1, F_2\}$			
		B_4	$\{F_1, F_2\}$			
		B_5	$\{F_1, F_2\}$	B_1	F_1	(1, 3)
$\tilde{G}_2$	3	B_2	$\{F_2, F_3\}$			
		B_3	$\{F_2, F_3\}$			
		B_4	$\{F_2, F_3\}$			
		B_5	$\{F_2\}$	B_5	F_2	(2, 7, 5)
$\tilde{G}_3$	4	B_2	$\{F_3\}$			
		B_3	$\{F_3, F_6\}$			
		B_4	$\{F_3\}$			
		B_6	$\{F_5\}$			
		B_7	$\{F_5, F_6\}$	B_2	F_3	(1, 4)
$\tilde{G}_4$	5	B_3	$\{F_6\}$			
		B_4	$\{F_7\}$			
		B_6	$\{F_5\}$			
		B_7	$\{F_5, F_6\}$	B_3	F_6	(3, 5)
$\tilde{G}_5$	6	B_4	$\{F_7\}$			
		B_6	$\{F_5\}$			
		B_7	$\{F_5, F_9\}$	B_4	F_7	(4, 6)
$\tilde{G}_6$	7	B_6	$\{F_5\}$			
		B_7	$\{F_5, F_9\}$	B_6	F_5	(6, 7)
$\tilde{G}_7$	8	B_7	$\{F_9\}$	B_7	F_9	(2, 8, 5)
$\tilde{G}_8$	9	B_8	$\{F_{15}\}$	B_8	F_{15}	(7, 8)
$\tilde{G}_9$	10	$(\lvert E\rvert - n + 2) = 10 = f$: algorithm terminates				

Bridge definitions

$B_1 = [(1, 3)]$, $B_2 = [(1, 4)]$, $B_3 = [(3, 5)]$
$B_4 = [(4, 6)]$
$B_5 = [(7, 8), (7, 2), (7, 5), (7, 6), (8, 2), (8, 5)]$
$B_6 = [(6, 7)]$, $B_7 = [(8, 2), (8, 5), (8, 7)]$
$B_8 = [(7, 8)]$

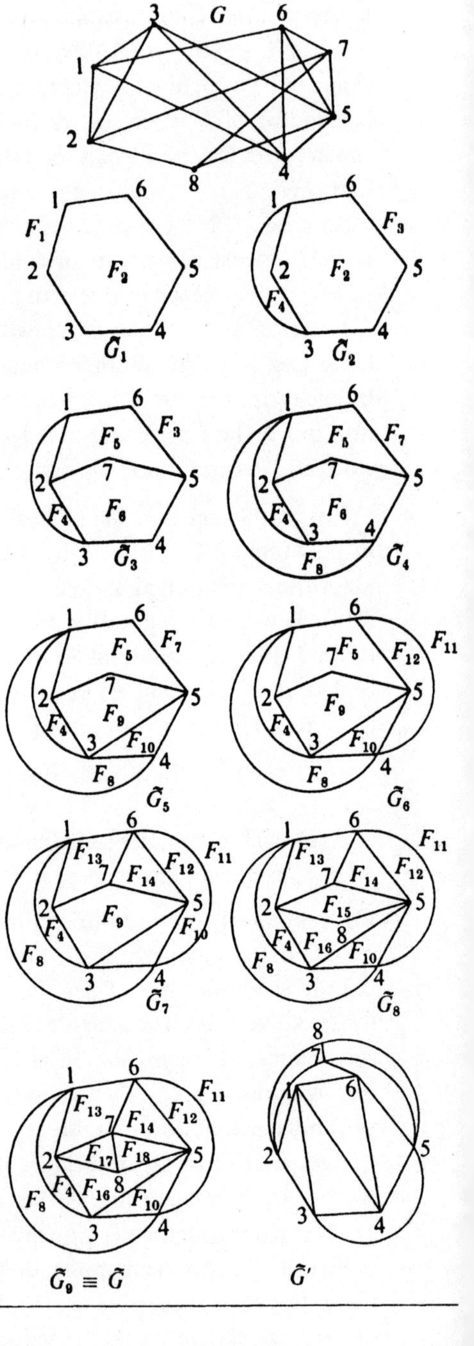

statement 13 of the algorithm and P_i as defined in statement 14. There is a separate table defining each bridge by its edge-set. As can be seen, in this case the algorithm terminates when $f = (|E|-n+2)$ with a planar embedding of G, $\tilde{G}_9$ and the message 'G is planar' would be output. The additional sketch labelled $\tilde{G}'$ represents a planar embedding of G which could have resulted if in going from $\tilde{G}_1$ to $\tilde{G}_2$ the path (1, 3) had been placed in F_2 rather than in F_1. This illustrates a point in the verification of theorem 3.10. Because G is planar, the bridges relative to $\tilde{G}_1$ that are finally placed in F_1 could all have been placed in F_2 and vice versa. This is rather a special example because $\tilde{G}'$ is not distinctly different from $\tilde{G}_9$. In fact, $\tilde{G}'$ can be obtained from $\tilde{G}_9$ merely by causing (see theorem 3.2) the face (2, 8, 5, 3) to become the exterior face. In general, however, given a choice of B and F as defined in statement 13 of the algorithm, distinctly different embeddings can be obtained.

Fig. 3.21. An application of the planarity testing algorithm.

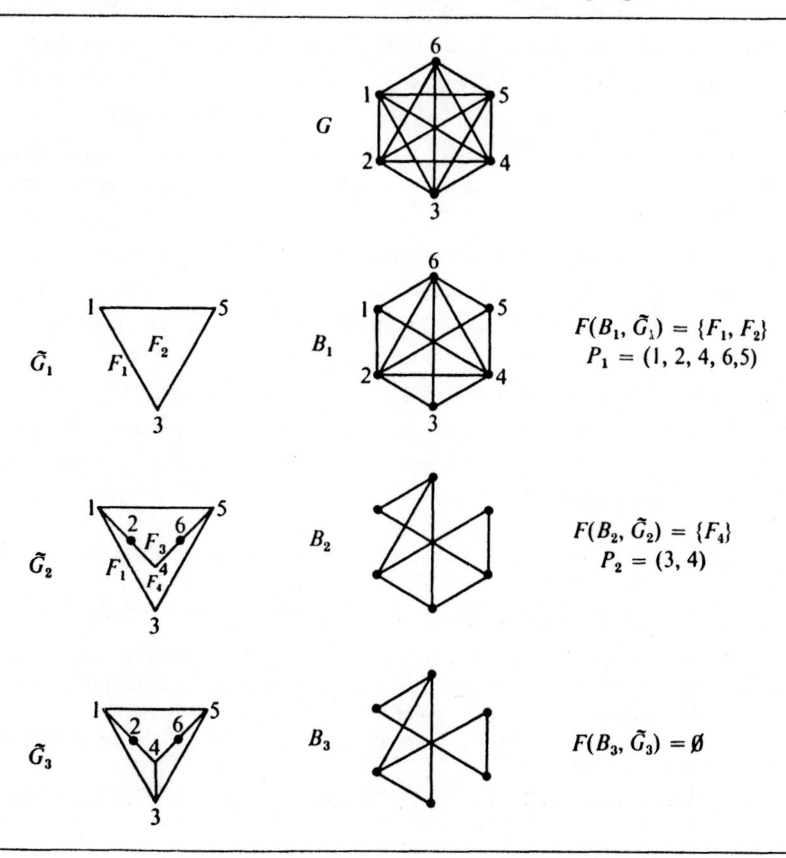

Finally, Figure 3.21 shows an application of the algorithm to the non-planar graph K_6. For each $\tilde{G}_i$ there is one bridge denoted by B_i, $F(B_i, \tilde{G}_i)$ and P_i also indicated in each case. The algorithm terminates when $F(B_3, \tilde{G}_3) = \varnothing$ with the message 'G is non-planar'.

3.5 Summary and references

Euler's formula provides a simple basis for deriving many immediate results relating to planar graphs. Some of the problems that follow provide further illustration of this. We also provided the extension to non-planar surfaces in section 3.2. The treatment of non-planar surfaces was informal, being illustrative rather than rigorous. Results in this area are highly specific and not of much practical benefit. Chapter 2 of Beineke & Wilson[7] provides a good commentary and selection of results. Chapter 11 of Harary[8] is also worthy of a reference.

The main characterisations of planarity we described were those of Kuratowski[9] and of Whitney[10] who used the idea of combinatorial dual. Our proofs of the relevant theorems are not based upon the original papers but on simpler expositions. The proof of theorem 3.5 is largely based on one given by Berge,[11] whilst the proof of theorem 3.8 is based on Parsons'.[12] Another well-known characterisation of planarity not covered in the text is that due to McLane[13]: a graph is planar if and only if it has a circuit basis (see section 2.2.1), together with one additional circuit such that this collection of circuits contains each edge of the graph twice. Finally, theorem 3.6 is essentially taken from Demoucron *et al.*[6]

A survey of early planarity testing algorithms is provided by Shirey.[14] As was stated earlier, linear time algorithms have been described by Hopcroft & Tarjan[2] and by Lempel *et al.*[3] Both of these algorithms receive detailed description in Even.[15] Our validification in theorem 3.10 of the planarity testing algorithm of Demoucron *et al.*,[6] which is rather simpler than that to be found in the original text, was influenced by the presentation of Bondy & Murty in [16].

[1] Lipton, R. J. & Tarjan, R. E. 'Applications of a planar separator theorem', *SIAM J. Comput.*, **9** (3) (1980).
[2] Hopcroft, J. & Tarjan, R. E. 'Efficient planarity testing', *JACM*, **21** (4), 549–68 (1974).
[3] Lempel, A., Even, S. & Cederbaum, I. 'An algorithm for planarity testing of graphs', *Theory of Graphs, International Symposium*, Rome 1966, P. Rosenstiehl, ed., Gordon and Breach, NY, pp. 215–32 (1967).
[4] Even, S. & Tarjan, R. E. 'Computing on st-numbering', *Theory of Computer Science*, **2**, 339–44 (1976).
[5] Lueker, G. S. & Booth, K. S. 'Testing for the consecutive ones property, interval graphs and graph planarity using *PQ*-tree algorithms', *J. of Comp. and Sys. Sciences*, **13**, 335–79 (1976).

[6] Demoucron, G., Malgrange, Y. & Pertuiset, R. 'Graphes planaires: reconnaissance et construction de représentations planaires topologiques', *Rev. Francaise Recherche Opérationnelle*, **8**, 33–47 (1964).
[7] Beineke, L. W. & Wilson, R. J. (eds.), *Selected Topics in Graph Theory*, chapter 2 (White & Lowell). Academic Press (1978).
[8] Harary, F. *Graph Theory*. Addison-Wesley (1969).
[9] Kuratowski, G. 'Sur le problème des courbes gauches en topologie', *Fund. Math.*, **15**, 271–83 (1930).
[10] Whitney, H. (*a*) 'Non-separable and planar graphs', *Trans. Am. Maths. Soc.*, **34**, 339–62 (1932).
Whitney, H. (*b*) 'Planar graphs', *Fund. Maths.*, **21**, 73–84 (1933).
Whitney, H. (*c*) '2-isomorphic graphs', *Am. J. Maths*, **55**, 245–54 (1933).
[11] Berge, C. *Theory of Graphs and its Applications*. John Wiley and Sons, NY (1962).
[12] Parsons, T. D. 'On planar graphs', *Am. Maths. Monthly*, **78** (2), 176–8 (1971).
[13] McLane, S. 'A combinatorial condition for planar graphs', *Fund Maths.*, **28**, 22–32 (1937).
[14] Shirey, R. W. Implementation and analysis of efficient planarity testing algorithms. PhD dissertation. Computer Sciences. University of Wisconsin (1969).
[15] Even, S. *Graph Algorithms*. Computer Science Press (1979).
[16] Bondy, J. A. & Murty, U. S. R. *Graph Theory with Applications*. The Macmillan Press (1976).
[17] Fáry, I. 'On straight line representation of planar graphs', *Acta. Sci. Math. Szeged.*, 229–33 (1948).
[18] Filotti, I. S., Miller, G. L. & Reif, J. 'On determining the genus of a graph in $O(V^{O(g)})$ steps', *Proceedings of the eleventh annual ACM Symposium on the Theory of Computing*. Atlanta, Georgia (1979).

EXERCISES

3.1. Given an arbitrary simple planar graph with n vertices and $|E|$ edges, show that the maximum number of edges, M, that can be added to the graph, subject to it remaining planar is given by

$$M = 3n - |E| - 6$$

(Use Euler's formula. When no more edges can be added every face of an embedding is triangular. Every simple planar graph is thus a subgraph of such a *planar triangulation*.)

3.2. Demonstrate that every simple graph with $|E| < 9$ or with $n < 5$ is planar.

3.3. (*a*) Three houses have to be connected individually to the sources of three amenities (electricity, gas and water). Show that this cannot be done without at least two of the lines of supply crossing.
(Because of this old problem, $K_{3,3}$ is sometimes known as the *amenities graph*.)

(*b*) Show that the Petersen graph (figure 6.14) contains a subgraph homeomorphic to $K_{3,3}$ and is therefore, according to Kuratowski's theorem, non-planar.

3.4. In a *completely regular* (simple planar) graph every vertex has the same degree $d(v)$, and every face has the same degree $d(f)$. Draw every completely regular (finite) graph. (For these graphs $2|E| = nd(v) = fd(f)$. Euler's formula then gives:

$$n = \frac{4d(f)}{2d(v)-d(f)\,(d(v)-2)}$$

For a fixed $d(v)$ we can find the allowable $d(f)$ consistent with a finite positive integer n. There are only five such graphs with $d(v) > 2$ and $d(f) > 2$.)

3.5. In the previous exercise we presumed that n was finite. Suppose, however, that $n = \infty$, then show that if G is completely regular and $d(v) > 2$ then $d(f)$ can only be 3, 4 or 6. This is a well-known fact in crystallography.

3.6. A *self-dual* is a simple planar graph which is isomorphic to its dual. Show, using Euler's formula, that if G is a self-dual then $2n = |E|+2$. How might a self dual be constructed for $n \geqslant 4$?
(Not every simple planar graph with $2n = |E|+2$ is a self-dual. Take care with vertices of degree 2.)

3.7. The *complement* $\bar{G}$ of a graph $G = (V, E)$ with n vertices is given by $\bar{G} = (K_n - E)$. Show that if $n \geqslant 11$, then at least one of G and $\bar{G}$ is non-planar.
(Use corollary 3.1. This result is also true for $n = 9$ and $n = 10$, but the proof is more difficult.)

3.8. Draw a planar embedding of the following graph in which every edge is a straight line.

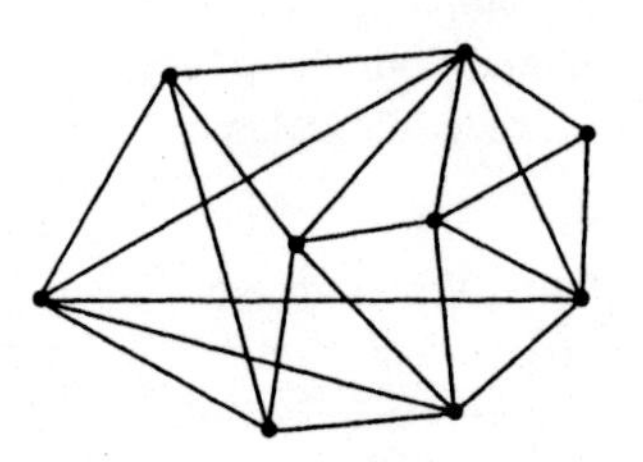

(Every simple planar graph has an embedding in which each edge is a straight line, Fáry[17].)

3.9. Show that the average degree of the vertices in a simple planar graph is less than 6 (in fact less than or equal to $[6-(12/n)]$). Thus provide a different proof from that in the text that any simple planar graph must have at least one vertex of degree at most 5.
(Use corollary 3.1 and that the average degree of the vertices is $2|E|/n$.)

3.10. Show that if G_1 is a dual of G_2 and that if G_1' is 2-isomorphic to G_1, then G_1' is also a dual of G_2.
(Establish first that there is a one-to-one correspondence between edges of G_1 and edges of G_1' and that a circuit in G_1 is a circuit in G_1' and vice-versa. This exercise proves one-half of theorem 3.7, proof of the other half is quite lengthy – see Whitney[10].)

3.11. An electrical circuit consists of connections between two sets of terminals A and B. Set A has six and set B has five terminals. Each member of A is connected to every member of B. Show by construction that such a circuit can be printed on two sides of an insulating sheet with terminals extending through the sheet.
[In general the thickness of a complete bipartite graph $K_{r,s}$ is given by (see the chapter by White & Lowell in [7]):

$$T = \left\lfloor \frac{rs}{2(r+s)-4} \right\rfloor$$

There may be some rare exceptions to this formula, but none has less than 48 vertices.]

3.12. Find three planar graphs such that their union is the complete graph on ten vertices, K_{10}.

3.13. Embed the complete graph on seven vertices, K_7, on a torus.

3.14. Describe the details of an implementation of the planarity testing algorithm of figure 3.17 which is as efficient as you can make it.

4

Networks and flows

A useful standpoint in solving a variety of problems is to model them in terms of some flow along the edges of a digraph. In some cases this flow may bear an obvious and direct analogy to the original problem. For others, flow may have been contrived to provide a novel or unexpected mode of solution.

This chapter provides an introduction to classical network flow theory. We describe an algorithm to maximise the flow across a suitably parameterised network and an algorithm to produce minimum-cost flows. Novel applications of this material may be found in the section on connectivity in this chapter, in the description of bipartite matching in chapter 5 and in the section on postman's tours in chapter 6.

4.1 Networks and flows

We start with some definitions. A (*transport*) *network* is a finite connected digraph in which:

(*a*) one vertex x, with $d^+(x) > 0$ is called the *source* of the network, and

(*b*) one vertex y, with $d^-(y) > 0$, is called the *sink* of the network.

A *flow* for the network N, associates a non-negative integer $f(u, v)$ with each edge (u, v) of N, such that for all vertices v, other than x or y:

$$\sum_u f(u, v) = \sum_u f(v, u)$$

Clearly, a network is a model for the flow of material leaving a single departure point (the source) and arriving at a single destination (the sink). Within the model $f(u, v)$ quantifies the flow along (u, v). The last equation ensures a conservation of flow at each vertex. In practice it is likely that there will be an upper bound on the possible flow along any edge. For each edge (u, v) this maximum, denoted by $c(u, v)$ and called the *capacity* of the edge, is a positive integer. For our purposes (although, see exercise 4.4) we take the minimum allowable flow along any edge to be zero. We now add to our definition of a flow by requiring that for each edge (u, v):

$$0 \leqslant f(u, v) \leqslant c(u, v)$$

A *cut* of a network $N = (V, E)$ is a cut-set of the underlying graph. The cut partitions V into two subsets P and $\bar{P}$ such that P contains x and $\bar{P}$ contains y. We denote the cut by $(P, \bar{P})$. Clearly, $P \cap \bar{P} = \varnothing$ and $P \cup \bar{P} = V$. The *capacity of a cut* $(P, \bar{P})$, denoted by $K(P, \bar{P})$ is defined to be the sum of the capacities of those edges incident from vertices in P and incident to vertices in $\bar{P}$:

$$K(P, \bar{P}) = \sum_{\substack{u \in P \\ v \in \bar{P}}} c(u, v)$$

The value of the flow $F(N)$ for a network $N = (V, E)$ is defined to be the net flow leaving the source x:

$$F(N) = \sum_{v} f(x, v) - \sum_{v} f(v, x)$$

We can now prove the following intuitively obvious theorem:

Theorem 4.1. For an arbitrary cut $(P, \bar{P})$ of the network N, the value of the flow is given by:

$$F(N) = \sum_{\substack{u \in P \\ v \in \bar{P}}} f(u, v) - \sum_{\substack{u \in \bar{P} \\ v \in P}} f(u, v)$$

$$= (\text{flow from } P \text{ to } \bar{P}) - (\text{flow from } \bar{P} \text{ to } P)$$

Proof. By definition

$$F(N) = \sum_{v} f(x, v) - \sum_{v} f(v, x)$$

Also, for any vertex $u \in P$ other than x:

$$0 = \sum_{v} f(u, v) - \sum_{v} f(v, u)$$

Summing these equations over all $u \in P$, including x, we obtain:

$$F(N) = \sum_{u \in P} \left(\sum_{v} f(u, v) - \sum_{v} f(v, u)\right)$$

Now:

$$\sum_{u \in P} \sum_{v} f(u, v) = \sum_{\substack{u \in P \\ v \in P}} f(u, v) + \sum_{\substack{u \in P \\ v \in \bar{P}}} f(u, v)$$

and

$$\sum_{u \in P} \sum_{v} f(v, u) = \sum_{\substack{u \in P \\ v \in P}} f(v, u) + \sum_{\substack{u \in P \\ v \in \bar{P}}} f(v, u)$$

Clearly, $\sum_{\substack{u \in P \\ v \in P}} f(v, u)$ is the same as $\sum_{\substack{u \in P \\ v \in P}} f(u, v)$ and so the theorem follows by substitution into the expression for $F(N)$. ■

Corollary 4.1. The value of the flow for any network cannot exceed the capacity of any cut $(P, \bar{P})$:

$$F(N) \leqslant \min (K(P, \bar{P}))$$

Proof. This follows directly from the previous theorem, since for any cut $(P, \bar{P})$:

$$\begin{aligned} F(N) &= \sum_{\substack{u \in P \\ v \in \bar{P}}} f(u, v) - \sum_{\substack{u \in \bar{P} \\ v \in P}} f(u, v) \\ &\leqslant \sum_{\substack{u \in P \\ v \in \bar{P}}} c(u, v) - \sum_{\substack{u \in \bar{P} \\ v \in P}} f(u, v) \\ &= K(P, \bar{P}) - \sum_{\substack{u \in \bar{P} \\ v \in P}} f(u, v) \\ &\leqslant K(P, \bar{P}) \end{aligned}$$

■

4.2 Maximising the flow in a network

Corollary 4.1 provides an upper bound for the *maximum-flow problem* which we now consider. The problem is simply to find a flow of maximum value in any given network.

A *path* Q from the source x to the sink y of a network $N = (V, E)$ is defined to be a sequence of distinct vertices $Q = (v_0, v_1, \ldots, v_k)$, where $v_0 = x$ and $v_k = y$ such that Q is a path from x to y in the underlying graph of N. Clearly, for any two consecutive vertices v_i and v_{i+1} of Q, either $(v_i, v_{i+1}) \in E$ or $(v_{i+1}, v_i) \in E$. In the former case (v_i, v_{i+1}) is called a *forward*-edge whilst in the latter case it is called a *reverse*-edge.

For a given flow $F(N)$ of N, a (*flow*) *augmenting path* is a path Q of N such that for each $(v_i, v_{i+1}) \in Q$:

(*a*) if (v_i, v_{i+1}) is a forward-edge then:

$$\Delta_i = c(v_i, v_{i+1}) - f(v_i, v_{i+1}) > 0$$

and

(*b*) if (v_i, v_{i+1}) is a reverse-edge then:

$$\Delta_i = f(v_{i+1}, v_i) > 0$$

If Q is an augmenting path then we define Δ as follows:

$$\Delta = \min \Delta_i > 0$$

Each (v_i, v_{i+1}) of Q for which $\Delta_i = \Delta$ is called a *bottleneck-edge* relative to $F(N)$ and Q.

For a given N and $F(N)$, if an augmenting path Q exists, then we can construct a new flow $F'(N)$ such that the value of $F'(N)$ is equal to the value of $F(N)$ plus Δ. We do this by changing the flow for each (v_i, v_{i+1}) of Q as follows:

(*a*) if (v_i, v_{i+1}) is a forward-edge then

$$f(v_i, v_{i+1}) \leftarrow f(v_i, v_{i+1}) + \Delta$$

and

(*b*) if (v_i, v_{i+1}) is a reverse edge then

$$f(v_{i+1}, v_i) \leftarrow f(v_{i+1}, v_i) - \Delta$$

Clearly, these changes preserve the conservation of flow requirement at each vertex excluding x and y so that $F'(N)$ is indeed a feasible flow. Moreover, the net flow from x is increased by the addition of Δ to the flow along (x, v_1).

Figure 4.1 shows a network in which each edge (u, v) is labelled with the pair $f(u, v)$, $c(u, v)$. Q is an augmenting path for which (x, v_1) and (v_3, y) are forward-edges, while (v_1, v_2) and (v_2, v_3) are reverse-edges. Each edge of the path except (v_3, y) is a bottleneck-edge and $\Delta = 1$. We can therefore augment the flow by making the following assignments:

$$f(x, v_1) \leftarrow 2, f(v_1, v_2) \leftarrow 0, f(v_2, v_3) \leftarrow 0, f(v_3, y) \leftarrow 2$$

Fig. 4.1

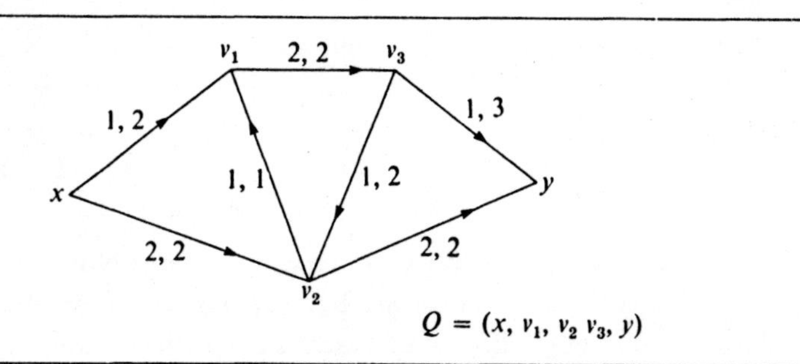

The idea of an augmenting path forms the basis of an algorithm, originally due to Ford & Fulkerson, for solving the maximum-flow problem. Starting from some initial flow $F_0(N)$, which could be the zero flow (i.e., $f(u, v) = 0$ for all $(u, v) \in E$), we construct a sequence of flows $F_1(N)$, $F_2(N)$, $F_3(N)$ .... $F_{i+1}(N)$ is constructed from $F_i(N)$ by finding a flow augmenting path along which $F_i(N)$ is augmented. Ignoring for the moment the question of finding augmenting paths, we can see that such an algorithm would work because:

(*a*) termination is guaranteed by observing that for all i the value of $F_{i+1}(N)$ is greater than the value of $F_i(N)$ and corollary 4.1 provides an upper bound on the maximum flow, and

(*b*) theorem 4.2 guarantees that if no augmenting path exists for $F_i(N)$, then $F_i(N)$ has a maximum value.

Theorem 4.2. If no augmenting path exists for some $F(N)$, then the value of $F(N)$ is a maximum.

Proof. We first describe a labelling process for the vertices of N. Initially no vertex is labelled and the labelling proceeds as follows:

(*a*) x is labelled.
(*b*) If for $(u, v) \in E$, u is labelled and v is unlabelled, then, provided $f(u, v) < c(u, v)$, v is labelled.
(*c*) If for $(u, v) \in E$, v is labelled and u is unlabelled, then, provided $f(u, v) > 0$, u is labelled.

By a repetition of (*b*) and (*c*) as many vertices of N as possible are labelled. It is easy to see that the above process cannot cause y to be labelled if no augmenting path exists. The labelling process defines a cut $(P, \bar{P})$ of N such that any labelled vertex is a member of P and any unlabelled vertex is a member of $\bar{P}$. From the labelling rules we deduce that:

$$f(u, v) = c(u, v) \quad \text{if} \quad u \in P \text{ and } v \in \bar{P}$$

$$f(u, v) = 0 \quad \text{if} \quad u \in \bar{P} \text{ and } v \in P$$

Thus, using theorem 4.1:

$$F(N) = \sum_{\substack{u \in P \\ v \in \bar{P}}} f(u, v) - \sum_{\substack{u \in \bar{P} \\ v \in P}} f(u, v) = \sum_{\substack{u \in P \\ v \in \bar{P}}} c(u, v) = K(P, \bar{P})$$

and so, by corollary 4.1, $F(N)$ must be a maximum. Notice that $(P, \bar{P})$ must be a cut of minimum capacity because if a cut of smaller capacity existed then the value of $F(N)$ would exceed that capacity and corollary 4.1 would be violated. ■

The algorithm we have outlined for the maximum-flow problem shows that it is always possible to attain a flow value $F(N)$ equal to min $(K(P, \bar{P}))$. This proves, along with corollary 4.1, the well-known *max-flow, min-cut* theorem originally stated by Ford & Fulkerson[1]:

Theorem 4.3 (max-flow, min-cut). For a given network the maximum possible value of the flow is equal to the minimum capacity of all cuts:

$$\max F(N) = \min K(P, \bar{P}).$$

Until now we have deliberately put to one side the question of how best to find an augmenting path at each step of the maximum-flow algorithm

This is because the question bears significantly upon the efficiency of the algorithm. If each augmentation only increases the overall flow from x to y by one unit, then the number of augmentations required for maximisation could be equal to min $K(P, \bar{P})$. This can be arbitrarily large and bear no relationship to the size of the network. For example, consider the network of figure 4.2. The edge (v_1, v_2) has a unit capacity whilst every other edge has a capacity of I. Starting with the zero flow, we could carry out a

Fig. 4.2

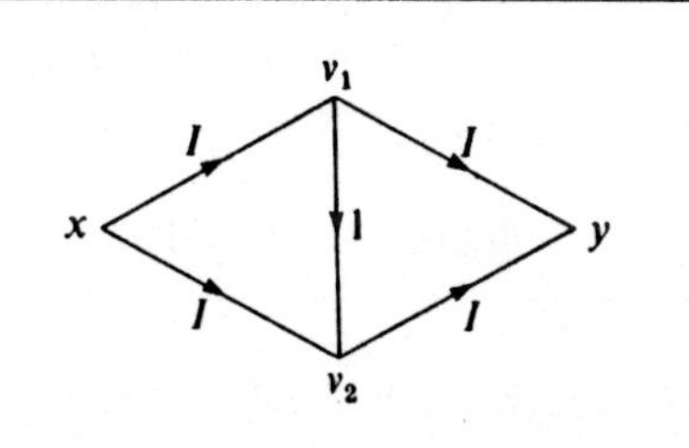

succession of augmentations alternately using the paths $P_1 = (x, v_1, v_2, y)$ and $P_2 = (x, v_2, v_1, y)$. Each augmentation enhances the flow by only one unit so that, overall, $2I$ augmentations will be required. We shall describe an algorithm of Edmonds & Karp[2] which chooses augmentation paths in such a way that the complete algorithm for flow maximisation operates in $O(n|E|^2)$-time. In fact more efficient algorithms are known (see, for example, Karzanov[3] and Malhotra *et al.*[4] for $O(n^3)$-algorithms), but they require considerably more explanation. We content ourselves with a demonstration that the maximum-flow problem can be solved in a time which is polynomially dependent upon the network size only. In other words, the complexity is independent of the edge capacities. We closely follow the work of Edmonds & Karp.

In fact, the following method of choosing augmentation paths due to Edmonds & Karp is so natural that it is likely to be included unwittingly in any implementation of the maximum-flow algorithm. Given a network $N = (V, E)$ with a flow F, we first construct an associated network $N^F = (V, E')$ such that there is a one-to-one correspondence between augmentation paths in N for F and directed paths from x to y in N^F. This is clearly the case if N and N^F have the same vertex set and if, for any two vertices u and v, (u, v) is an edge of N^F if and only if either:

$$(u, v) \in E \quad \text{and} \quad c(u, v) - f(u, v) > 0$$

or

$$(v, u) \in E \quad \text{and} \quad f(v, u) > 0$$

Thus the question of finding an augmentation path for N and F is reduced to finding a directed path from x to y in N^F. We denote a directed path in N^F by P^F and its corresponding path for $F(N)$ by P. Any edge of P^F corresponding to a bottleneck-edge of $F(N)$ is also called a bottleneck-edge. We now have to describe the precise method by which P^F is found.

In determining P^F each vertex v of N^F is first labelled $L(v)$, where $L(v)$ is equal to the minimum distance (in edges) from x to v. If no such path exists then $L(v) = 0$. This can be done using the breadth-first search for shortest paths algorithms which is detailed in exercise 1.15. If a path exists from x to y, then P^F is chosen to be a path of minimum length. This can be traced backwards from y by next visiting a vertex u such that $L(u) = L(v) - 1$, where v is the current vertex being visited.

Fig. 4.3. The breadth-first search for P^K procedure, *BFSPK*.

```
1.  begin
    Carry out a breadth-first search for the shortest distances in N^F
              from x to each vertex v. (L(v) > 0, if v ≠ x, denotes
              this path length and if L(v) = 0 then no path exists.)
2.  if L(y) = 0 then PATH ← false
        else
        begin
3.      for all v ∈ V construct B'(v)
4.      P^F ← (y)
5.      u ← y
6.      while u ≠ x do
              begin
7.            find a vertex v such that v ∈ B'(u) and L(u) = L(v)+1
8.            add v to the head of P^F
9.            u ← v
              end
        end
    end
```

Figure 4.3 encapsulates this breadth-first search procedure for P^K within a procedure called *BFSPK*. Line 1 represents the labelling process detailed in exercise 1.15, which we recall operates in $O(|E|)$-time. Line 2 simply determines whether or not a path P^F exists. If not, a boolean variable *PATH* is assigned the value **false** and the procedure terminates. If a path P^F does exist then the rest of the procedure is designed to construct the list of vertices defining it. We can do this efficiently by making available for each vertex v, a list $B'(v)$. This contains those vertices $u_1, u_2, \ldots$, such that $(u_1, v), (u_2, v), \ldots$ are edges of N^F. In other words, $B'(v)$ specifies all the edges incident *to* v, as opposed to the adjacency list $B(v)$ which specifies all those edges incident *from* v. We imagine that the adjacency lists are available globally, then it is easy to see that line 3 can construct all the $B'(v)$ in $O(|E|)$-time. The benefit afforded by the $B'(v)$ can be seen within the **while** statement beginning at line 6. This constructs P^F by adding the current vertex being visited to the head of the list of vertices defining that portion of P^F that has been traced so far from y. In line 7, the use of the $B'(u)$ means that the search for v need only inspect $d^-(u)$ vertices. Thus for all iterations of the body of the while statement, these searches require at worst $\sum_u d^-(u) = O(|E|)$ inspections. We conclude that the breadth-first search for P^K procedure executes in $O(|E|)$-time. In order to establish the overall complexity of the maximum-flow algorithm, using shortest augmentation paths, we require the following theorem.

Theorem 4.4. If in the maximum-flow algorithm each augmentation is carried out along a shortest path then a maximum flow will be obtained after no more than $\frac{1}{2}|E|\cdot n$ augmentations.

Proof. Let $F^0, F^1, F^2, \ldots$ be a sequence of flows in N such that F^{k+1} is obtained from F^k by an augmentation corresponding to a shortest path P^k in N^{F^k}. We shall write N^k for N^{F^k} and denote by $d^k(u, v)$ the shortest distance from u to v in N^k. If no path exists from u to v then we take $d^k(u, v) = \infty$.

In order to proceed with the proof we need two lemmas:

Lemma 4.1. If $k < m$ and (u, v) is a bottleneck-edge relative to P^k and F^k, and also relative to P^m and F^m, then for some l such that $k < l < m$, $(v, u) \in P^l$.

Clearly, if (u, v) is a bottleneck-edge relative to P^k and F^k, then it will not be an edge of N^{k+1}. It can only be reintroduced as an edge for some subsequent N^m if the flow from u to v along (u, v) is reduced in some intermediate augmentation, say for N^l. This is only possible if $(v, u) \in P$.

Lemma 4.2. If $k < l, (u, v) \in P^k$ and $(v, u) \in P^l$, then $d^l(x, y) \geqslant d^k(x, y)+2$.

In order to prove this lemma we first need to show that for all k and u:

$$d^k(x, u) \leqslant d^{k+1}(x, u)$$

and

$$d^k(u, y) \leqslant d^{k+1}(u, y)$$

We shall first prove the first statement, proof of the second being very similar. If $d^{k+1}(x, u) = \infty$ then the result is trivial. If $d^{k+1}(x, u)$ is finite then we denote a shortest path from x to u in N^{k+1} by

$$(u_0 = x, u_1, u_2, \ldots, u_h = u)$$

Now $d^k(x, u_0) = 1$ and

$$d^k(x, u_{i+1}) \leqslant 1+d^k(x, u_i), \quad i = 0, 1, \ldots, (h-1)$$

because since $(u_i, u_{i+1}) \in N^{k+1}$ then either $(u_i, u_{i+1}) \in N^k$ or $(u_{i+1}, u_i) \in P^k$. In the former case $d^k(x, u_{i+1}) \leqslant 1+d^k(x, u_i)$ since (u_i, u_{i+1}) enables us to get a directed path from x to u_{i+1} in N^k having no more than $1+d^k(x, u_i)$ edges. In the latter case

$$d^k(x, u_i) = 1+d^k(x, u_{i+1})$$

so

$$d^k(x, u_{i+1}) = -1+d^k(x, u_i) < 1+d^k(x, u_i)$$

Summing the set of inequalities over all i, we obtain:

$$\sum_{i=0}^{h-1} d^k(x, u_{i+1}) \leqslant h + \sum_{i=0}^{h-1} d^k(x, u_i)$$

so that

$$d^k(x, u) \leqslant h+d^k(x, x) = d^{k+1}(x, u)$$

which is the inequality we set out to prove.

We can now complete the proof of lemma 4.2. Since $(u, v) \in P^k$:

$$d^k(x, y) = d^k(x, u)+1+d^k(v, y)$$

and

$$d^k(x, v) = 1+d^k(x, u)$$

$$d^k(u, y) = 1+d^k(v, y)$$

Also $(v, u) \in P^l$ so that

$$d^l(x, y) = d^l(x, v)+1+d^l(u, y)$$

and using the inequalities previously obtained it follows that:

$$\begin{aligned} d^l(x, y) &\geqslant d^k(x, v)+1+d^k(u, y) \\ &= (1+d^k(x, u))+1+(1+d^k(v, y)) \\ &= 2+(d^k(x, u)+1+d^k(v, y)) \\ &= 2+d^k(x, y) \end{aligned}$$

Hence lemma 4.2 is proved.

Given these two lemmas the proof of theorem 4.4 is easily obtained. Let u and v be two vertices such that either (u, v) or (v, u) is an edge of N. The sequence $\{K_i\}$ consists of those indices K_i such that either (u, v) or (v, u) is a bottleneck-edge relative to P^{K_i} and F^{K_i}. Utilising lemma 4.1 we can find a sequence $\{l_j\}$, with $\{K_i\}$ as a subsequence such that:

$$(u, v) \in P^{l_j}, \quad j \text{ odd and } (v, u) \in P^{l_j}, j \text{ even}$$

or

$$(u, v) \in P^{l_j}, \quad j \text{ even and } (v, u) \in P^{l_j}, j \text{ odd}$$

Hence, by lemma 4.2:

$$d^{l_{j+1}}(x, y) \geqslant d^{l_j}(x, y)+2$$

so that in consecutive appearances of (u, v) or (v, u) as a bottleneck, the length of the augmentation path increases by at least two edges. Since an augmentation path cannot be less than one edge or more than $(n-1)$ edges in length, it follows that any edge of the network cannot provide a bottleneck more than $\frac{1}{2}n$ times. Every augmentation path contains at least one bottleneck and there are $|E|$ edges in the network. It follows that in the maximum-flow algorithm no more than $\frac{1}{2}|E| \cdot n$ shortest path augmentations will be required. ■

The breadth-first search for P^K procedure *BFSPK* of figure 4.3 is called in line 5 of the maximum-flow algorithm outlined in figure 4.4. Using theorem 4.4, we can now see that the complexity of this maximum-flow

Fig. 4.4. The maximum-flow algorithm.

---

1. Input the adjacency lists $A(v)$, edge capacities and initial edge
        flows for the network $N = (V, E)$
2. *PATH* ← **true**
3. **while** *PATH* = **true do**
        begin
4.   Construct the adjacency lists $B(v)$ for N^K, for each edge
                (u, v) recording $\Delta(u, v)$ and whether (u, v) is a forward
                or reverse edge.
5.   *BFSPK*
6.   **if** *PATH* = **true then**
                begin
7.   find $\Delta = \min \Delta(u, v), (u, v) \in P^K$
8.   **for all** $(u, v) \in P^K$ **do**
9.   **if** (u, v) is a forward edge of P^K
10.   **then** $f(u, v)+\Delta$
11.   **else** $f(v, u) \leftarrow f(v, u)-\Delta$
                end
        end

---

algorithm is $O(n|E|^2)$. Line 1 of the algorithm merely inputs the network N in adjacency list description and so requires $O(|E|)$-time. One convenience that might be adopted here is to append to each appearance of a vertex $u \in A(v)$ a record of $c(u, v)$ and $f(u, v)$. Line 2 initialises the boolean global *PATH*, which records whether or not an augmentation path exists for the current N^K. Each iteration of the while statement corresponds to one augmentation of the flow in N. In line 4 the adjacency lists $B(v)$ for the current N^K are constructed. This requires $O(|E|)$-time. For each $(u, v) \in E$ we add v to $B(u)$ if $c(u, v) > f(u, v)$, recording within appended locations that (u, v) is a forward-edge and that $\Delta(u, v) = c(u, v)-f(u, v)$, and if

$f(u, v) > 0$ then u is added to $B(v)$ noting that (v, u) is then a reverse-edge and that $\Delta(v, u) = f(v, u)$. Line 5 then determines the augmentation path P^K within $O(|E|)$ steps as previously described. If no path exists then, within the call of *BFSPK*, there is an assignment of **false** to *PATH* and the computation stops. If an augmentation path does exist then the conditional statement starting at line 6 augments the flow in N as previously described. Since there are at most $(n-1)$ edges in P^K, this requires $O(n)$ steps. Altogether, the body of the while statement therefore requires $O(|E|)$ operations. According to theorem 4.4 at most $\frac{1}{2}|E|\cdot n$ iterations will be required. It follows that the complete algorithm of figure 4.4 has $O(n|E|^2)$-complexity.

4.3 Menger's theorems and connectivity

As anticipated in section 2.2.3 we prove here some well-known theorems of Menger. In doing so we make use of the max-flow, min-cut theorem. In a natural way these theorems furnish us with algorithms to determine the vertex- and edge-connectivities of a graph. We shall be describing these also. First we require some definitions.

By $p_e(v_i, v_j)$ we denote the maximum number of edge disjoint paths between v_i and v_j. Similarly, by $p_v(v_i, v_j)$ we denote the maximum number of vertex (other than v_i and v_j) disjoint paths from v_i to v_j. By $c_e(v_i, v_j)$ we denote the smallest cardinality of those cut-sets which partition the graph so that v_i is in one component and v_j is in the other. Also, we define $c_v(v_i, v_j)$ to be the smallest cardinality of those vertex-cuts which separate G into two components, one containing v_i and the other containing v_j. Clearly, no such vertex-cut exists if (v_i, v_j) is an edge of the graph.

The following is a variation of one of Menger's theorems.

Theorem 4.5. Let $G = (V, E)$ be an undirected graph with $v_i, v_j \in V$, then $c_e(v_i, v_j) = p_e(v_i, v_j)$.

Proof. From G we construct a network N as follows. N contains the same vertex-set as G and for each edge (u, v) of G, N contains the directed edges (u, v) and (v, u). For each edge e of N, we assign a capacity $c(e) = 1$. Thus for any flow in N, $f(e) = 0$ or 1. We denote the maximum value of a flow from a source x to a sink y of N by F.

We first show that $F = p_e(x, y)$. If there are $p_e(x, y)$ edge disjoint paths from x to y in G, then there are $p_e(x, y)$ edge disjoint paths from x to y in N. Each such path can be used to transport one unit of flow from x to y. Thus $F \geqslant p_e(x, y)$. For a maximal flow from x to y in N we can, without loss of generality, assume that for each edge (u, v) not both of $f(u, v)$ and $f(v, u)$ are unity. If they were, then we could replace each flow by zero without affecting F. Flow can then be considered to consist of unit flows from x to y in N, corresponding to *edge disjoint* paths in G. Thus $F \leqslant p_e(x, y)$, and using our first result this yields $F = p_e(x, y)$.

According to the max-flow, min-cut theorem F is equal to the capacity of a cut-set $C = (P', P'')$ of N, $x \in P'$ and $y \in P''$. Every path from x to y in N uses at least one edge of C. The corresponding set of paths in G each uses at least one edge (u, v) such that (u, v) of N is in C. This set of edges will disconnect G and so we have a cut-set with cardinality F. Thus

$$C_e(x, y) \leqslant F = p_e(x, y)$$

We can easily see that $C_e(x, y) \geqslant p_e(x, y)$, because every path from x to y uses at least one edge of a set which disconnects G, and no two paths use the same edge. Thus $c_e(x, y) = p_e(x, y)$. ■

The following corollary is the more usual statement of Menger's edge-connectivity theorem.

Corollary 4.2. A graph is k-edge connected if and only if any two distinct vertices are connected by at least k-edge disjoint paths.

Proof. This follows directly from theorem 4.5 and the definition of a k-edge-connected graph (see section 2.2.3). ■

We are now in a position to describe a polynomial time algorithm to find the edge-connectivity $K_e(G)$ of an arbitrary graph $G = (V, E)$. From the definitions of both $K_e(G)$ and $c_e(v_i, v_j)$ it is evident that:

$$K_e(G) = \min_{v_i, v_j \in V} c_e(v_i, v_j)$$

We can therefore find $K_e(G)$ by solving the maximum-flow problem (perhaps using the algorithm of figure 4.4) for a series of networks each derived from G as in the proof of theorem 4.5. An immediate reaction might be that $O(n^2)$ maximisations are required because there are $n(n-1)$ different pairs of vertices. However, $O(n)$ maximisations will suffice. Notice that if for some network, $(P, \bar{P})$ is a cut-set of minimum cardinality, with v_i and v_j *any* two vertices such that $v_i \in P$ and $v_j \in \bar{P}$, then $K_e = c_e(v_i, v_j)$. It follows that K_e will be found by solving only those maximum-flow problems for which a particular vertex, say u, is the source. The remaining vertices are then taken as the sink in turn. Thus only $(n-1)$ maximisations are required.

Figure 4.5 outlines the algorithm for $K_e(G)$ which results from the above considerations. $\bar{G}$ denotes the digraph obtained by replacing each edge of G by two antiparallel edges, each of unit capacity. Line 3 simply assigns a convenient and suitably large initial value to K_e. The major and time consuming part of the algorithm is embodied in the **for** statement starting at line 4. Each application of line 5 finds the value of a maximum flow F for the network consisting of $\bar{G}$ with a source x and a sink y. If the algorithm of figure 4.4 is utilised then line 5 requires $O(n|E|^2)$-time. With $O(n)$ repetitions of this, we see that, overall, the algorithm of figure 4.5 would have $O(n^2|E|^2)$-complexity. This can, of course, be improved by using one of the more efficient maximum-flow algorithms referred to earlier.

Fig. 4.5. Algorithm to find the edge-connectivity $K_e(G)$ of an undirected graph G.

```
1.  Input G and constructs Ḡ
2.  Specify u
3.  K_e ← |E|
4.  for all v ∈ V−{u} do
        begin
5.      find F for (Ḡ with x = u and y = v)
6.      if F < K_e then K_e ← F
        end
7.  Output K_e
```

We turn to the problem of evaluating the vertex-connectivity $K_v(G)$ of a graph G. Our treatment is very similar to that for $K_e(G)$ but with minor complications. First we require a theorem analogous to theorem 4.5.

Theorem 4.6. If $(x, y) \notin E$ then $c_v(x, y) = p_v(x, y)$.

Proof. Given $G = (V, E)$, we construct a digraph $\bar{G}$ as follows. For every vertex $v \in V$, $\bar{G}$ contains two vertices v' and v'' and an edge (v', v'') called an *internal edge*. In addition, for every edge $(v_i, v_j) \in E$, $\bar{G}$ contains two edges (v_i'', v_j') and (v_j'', v_i') which we call *external edges*. We now define a network N consisting of the digraph $\bar{G}$ in which the source is x'' and the sink is y'. The capacity of each internal edge is one, and each external edge has an infinite capacity. Figure 4.6 shows N for the graph G illustrated. If we now denote the value of a maximum flow from x'' to y' by F, our proof proceeds like that for theorem 4.5.

We first show that $F = p_v(x, y)$. If there are $p_v(x, y)$ vertex disjoint paths in G, then we can identify $p_v(x, y)$ vertex disjoint paths from x'' to y' in $\bar{G}$. For this we simply associate with the path $(x, v_1, v_2, \ldots, y)$ in G, the path $(x'', v_1', v_1'', v_2', v_2'', \ldots, y)$ in $\bar{G}$. In $\bar{G}$ these $p_v(x, y)$ paths can be used to carry unit flows from x'' to y' so that $p_v(x, y) \leqslant F$. Now consider a flow in N. For each edge e, $f(e)$ is either zero or one. This is easily seen by noting that the flow through each vertex v is effectively bound by unity because either a single edge which has unit capacity is incident *to* v or a single edge which has unit capacity is incident *from* v. Thus any flow from x'' to y' can be decomposed into unit flows carried along *vertex disjoint* paths. These correspond to a set of vertex disjoint paths in G. Hence $p_v(x, y) \geqslant F$ and so we have completed the proof that $F = p_v(x, y)$.

Fig. 4.6

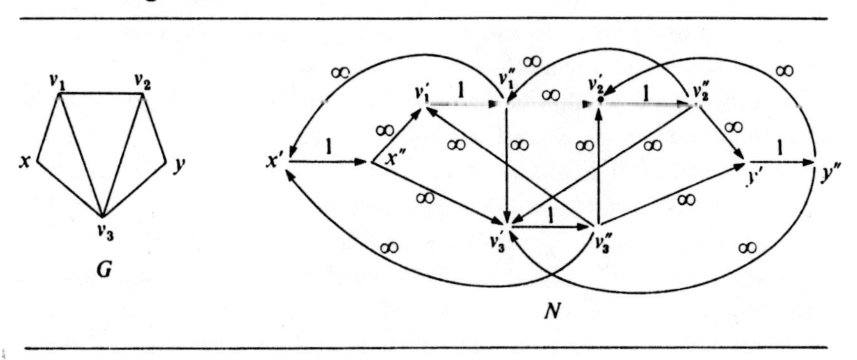

According to the max-flow, min-cut theorem, F is equal to the capacity of a cut-set $C = (P', P'')$ such that $x'' \in P'$ and $y' \in P''$. Moreover, every edge from P' to P'' must be an internal edge because the capacity of the cut, equal to F, is finite. Every path from the source to the sink in N uses at least one of these internal edges of C. Hence every path from x to y in G passes through a corresponding vertex. This set of vertices is therefore a vertex-cut of G, has cardinality F and is such that its removal from G produces two components one containing x and the other containing y. Hence $c_v(x, y)$ is at most F: $c_v(x, y) \leqslant F = p_v(x, y)$. Also $p_v(x, y)$ cannot exceed $c_v(x, y)$ because every one of the paths from x to y uses at least one vertex of a vertex-cut of size $c_v(x, y)$ and no two paths use the same vertex. Thus $c_v(x, y) = p_v(x, y)$. ■

The following corollary is Menger's vertex-connectivity theorem.

Corollary 4.3. A graph is k-vertex-connected if and only if any two distinct vertices are connected by at least k-(internally) vertex disjoint paths.

Proof. This follows directly from theorem 4.6 and the definition of a k-vertex-connected graph (see section 2.2.3). ■

From section 2.2.3, if G is complete then $K_v(G) = (n-1)$ and otherwise:

$$K_v(G) = \min_{(v_i, v_j) \notin E} c_v(v_i, v_j)$$

In view of theorem 4.6 and the definition of $K_v(K_n)$ we may therefore write, whether G is complete or otherwise:

$$K_v(G) = \min_{(v_i, v_j) \notin E} p_v(v_i, v_j)$$

and we can find $K_v(G)$ by solving the maximum-flow problem for a series of networks each derived from G as in the proof of theorem 4.6. Incidentally (see exercise 4.7), we can drop the requirement in the last equation that v_i and v_j have to be non-adjacent. If $\min p_v(v_i, v_j)$ occurs for two adjacent vertices, then it also occurs for two non-adjacent vertices.

If we base an algorithm to find $K_v(G)$ on the last equation, how many maximum-flow problems need to be solved? We can see that $O(|E|)$ are always sufficient. If a vertex-cut of minimum cardinality produces one component with a vertex-set V' and another component with a vertex-set V'', with v_i and v_j *any* two vertices such that $v_i \in V'$ and $v_j \in V''$, then $K_v = p_v(v_i, v_j)$. It follows that K_v will be found by solving a maximum-flow problem for which the source is in V' and the sink is in V''. Such a problem is guaranteed to be solved in the following process. First solve all those maximum-flow problems with v_1 as the source (taking in turn each of $v_j, j = 2, 3, \ldots, n$ as the sink, provided $(v_1, v_j) \notin E$), then those with v_2 as the source (taking in turn $v_j, j = 3, 4, \ldots, n$, as the sink, provided $(v_2, v_j) \notin E$) and so on until v_k has taken a turn as the source where $k = K_v(G)+1$. In this way one of $v_1, v_2, \ldots, v_k$, say v_i, is guaranteed not to be contained in a vertex-cut of size $K_v(G)$. The process solves all maximisation problems with v_i as source and so $K_v(G)$ will be found. For a given vertex as the source there are $O(n)$ possible sinks. Thus in all there are $O(K_v(G) \cdot n)$ maximum-flow problems to be solved. Using the following theorem we see that this is bound by $O(|E|)$ because, according to theorem 2.9, $K_v(G) \leqslant K_e(G)$.

Theorem 4.7. For $G = (V, E)$, $K_e(G) \leqslant 2|E|/n$.

Proof. From theorem 2.9, $K_e(G) \leqslant \delta$, however:

$$n.\delta \leqslant \sum_{v_i \in V} d(v_i) = 2|E|$$

and so the result follows. ■

Figure 4.7 outlines the algorithm for $K_v(G)$ based on the preceding considerations. Line 1 inputs G and constructs $\bar{G}$ defined in the proof of theorem 4.6 and exemplified in figure 4.6. $\bar{G}$ has $2n$ vertices and $(n+2|E|)$ edges. It can therefore be constructed ($|E| \geqslant n$) in $O(|E|)$-time. Line 2 initialises K_v to n and each subsequent assignment to K_v (in line 8) records the smallest value of $p_v(v_i, v_j)$ found so far. Each execution of the body of the **while** statement corresponds to solving all those maximum-flow problems with a given vertex v_i as the source x. F denotes the value of a maximum flow and y denotes the sink. Notice that on termination of the **while** statement, $i > K_v$. This means that the currently held value of the vertex-connectivity is less than the number of vertices which have been used as a source in the maximum-flow problems solved in line 7. This is

Fig. 4.7. Algorithm to find the vertex connectivity of $G = (V, E)$.

```
1. Input G and construct Ḡ
2. K_v ← n
3. i ← 0
4. while i ≯ K_v do
       begin
5.     i ← i+1
6.     for j = i+1 to n do
           begin
7.         if (v_i, v_j) ∉ E find F for (Ḡ with x = v_i and y = v_j)
8.         if F < K_v then K_v ← F
           end
       end
9. Output K_v
```

in accord with the previous discussion concerning the number of maximisations required to be solved. Line 7 is dominant as far as the complexity of the algorithm is concerned. If the algorithm of figure 4.4 is used to determine F, each maximisation takes $O(n|E|^2)$-time, so that, overall, K_v can be found in $O(n|E|^3)$-time. Of course it is easy to construct faster algorithms by utilising the faster algorithms to find F which we referred to earlier.

4.4 A minimum-cost flow algorithm

In section 4.1 we solved the problem of maximising the flow in a network in which each edge (u, v) had a maximum allowed flow of $c(u, v)$. In this section we wish to associate a further parameter $a(u, v)$ with each edge, where $a(u, v)$ is the *cost* of sending a unit of flow along (u, v). This has an obvious interpretation in any real transport network where the unit cost of transportation may vary from edge to edge depending upon the nature of that transport. This extra consideration of costs obviously poses new problems. In this section we consider just one, called the *minimum-cost flow problem*. This is the problem of how to transport V units of flow across a network such that a *minimum cost is incurred*. We describe here one method of solution, due to Ford & Fulkerson,[6] which is called the *minimum-cost flow algorithm.* The algorithm is in fact an interesting generalisation of the maximum-flow algorithm described earlier.

Each edge (u, v) of our network N has a maximum capacity $c(u, v)$ and an integer non-negative cost parameter $a(u, v)$. As before, we denote the source by x and the sink by y. The problem is to construct a minimum-cost feasible flow of V units. We can formulate this as a linear programming problem (see the appendix on linear programming) as follows:

$$\text{minimise} \sum_{(u,v)} a(u, v) f(u, v) \qquad \text{(i)}$$

subject to the constraints:

$$\sum_{v} (f(u, v) - f(v, u)) = 0, \text{ for all } u \neq x \text{ or } y \qquad \text{(ii)}$$

$$(\sum_{v} (f(x, v) - f(v, x)) - V) = 0 \qquad \text{(iii)}$$

$$(\sum_{v} (f(y, v) - f(v, y)) + V) = 0 \qquad \text{(iv)}$$

$$f(u, v) \leqslant c(u, v), \text{ for all } (u, v) \qquad \text{(v)}$$

and the non-negativity conditions:

$$f(u, v) \geqslant 0, \text{ for all } (u, v) \qquad \text{(vi)}$$

These statements have the following interpretation, (i) is the total cost of the flow, (ii) ensures that there is conservation of flow at all vertices other than x or y, (iii) and (iv) require that the flow from the source and into the sink is V while (v) and (vi) ensure that the edge flows are feasible.

Ford & Fulkerson solved the problem in a slightly different form, replacing (i) by the following:

$$\text{maximise } (pV - \sum_{(u,v)} a(u, v) f(u, v)) \qquad \text{(vii)}$$

The idea is to solve a sequence of linear programs, one for each consecutive value of the parameter $p = 0, 1, 2, \ldots$. V, the value of the network flow, is explicitly treated as a variable. For each linear program, p is the maximum cost that any unit of flow incurs in getting from x to y. Given a maximum flow at minimum cost for p, a maximum flow at minimum cost is then found for $(p+1)$. This is achieved by incrementing the flow along all possible paths which incur a total cost of $(p+1)$ in sending a unit of flow from x to y. The cost along such an augmenting path is calculated by summing the edge costs of forward-edges and subtracting the edge costs of reverse-edges.

Given our definition of p we can see that the value of expression (vii) can never be negative. This is because pV would be the cost of sending all units of flow from x to y, each unit at maximum cost p, while $\Sigma a(u, v) f(u, v)$ is the actual cost. Hence, maximising (vii) is equivalent to minimising (i).

Verification of the algorithm is possible through a consideration of the dual (see the appendix) of the pth problem described in (ii)–(vii). The dual is:

$$\text{minimise} \sum_{(u,v)} c(u, v)\gamma(u, v) \qquad \text{(i}'\text{)}$$

subject to the constraints:

$$-\pi(x)+\pi(y) = p \qquad \text{(ii}'\text{)}$$

$$\pi(u)-\pi(v)+\gamma(u, v) \geqslant -a(u, v), \text{ for all } (u, v) \qquad \text{(iii}'\text{)}$$

and the non-negativity conditions:

$$\gamma(u, v) \geqslant 0, \text{ for all } (u, v) \qquad \text{(iv}'\text{)}$$

while $\pi(u)$ is unconstrained in sign for all u. Here the dual variables $\pi(u)$ correspond to equations (ii), (iii) and (iv) while the dual variables $\gamma(u, v)$ correspond to the constraints (v). We call $\gamma(u, v)$ *edge numbers* and $\pi(u)$ *vertex numbers*. The equality sign in (ii′) appears because we have not explicitly restricted the sign of V, although it would be a trivial matter to do so. The complementary slackness conditions (see the appendix) are, for all (u, v):

$$(\pi(u)-\pi(v)+\gamma(u, v)) > -a(u, v) \Rightarrow f(u, v) = 0$$

and:

$$\gamma(u, v) > 0 \Rightarrow f(u, v) = c(u, v)$$

Hence, if we let

$$\pi(x) = 0, \pi(y) = p \qquad (a)$$

and

$$\gamma(u, v) = \max\{0, \pi(v)-\pi(u)-a(u, v)\} \qquad (b)$$

for all (u, v), then the slackness conditions become:

$$\pi(v)-\pi(u) < a(u, v) \Rightarrow f(u, v) = 0 \qquad (c)$$

and

$$\pi(v)-\pi(u) > a(u, v) \Rightarrow f(u, v) = c(u, v) \qquad (d)$$

Therefore, if we can find values for the vertex numbers and edge flows which satisfy (*a*), (*c*) and (*d*), and if the edge numbers are defined by (*b*), then we have found optimal solutions for the primal and dual problems for the current value of p.

As for the maximum-flow problem, the present algorithm uses a labelling process. This is as follows. Initially every vertex except the source is unlabelled. New vertices v are labelled if they are adjacent to one, u, that has already been labelled if either:

$$f(u, v) < c(u, v) \quad \text{and} \quad (\pi(v)-\pi(u)) = a(u, v)$$

or if:

$$f(u, v) > 0 \quad \text{and} \quad (\pi(v)-\pi(u)) = a(u, v)$$

Notice that if within this process the sink becomes labelled, then a flow augmenting path has been found. This is precisely as was the case for the maximum-flow algorithm but here we have the additional fact that the path cost is p. We can see this as follows. We define F to be the set of forward-edges of the path P and R to be the set of reverse-edges. Then we let P' be the set such that if $(u, v) \in F$ then $(u, v) \in P'$ and if $(v, u) \in R$, then $(u, v) \in P'$. It follows that:

$$\begin{aligned} \text{path cost} &= \sum_{(u,v)\in F} a(u, v) - \sum_{(v,u)\in R} a(v, u) \\ &= \sum_{(u,v)\in P'} (\pi(u) - \pi(v)) \\ &= \pi(x) + \pi(y) = p \end{aligned}$$

We can now provide a description of the minimum-cost flow algorithm outlined in figure 4.8. Lines 1 and 2 initialise the flows $f(u, v)$ and the vertex numbers $\pi(u)$ to the value of zero. The algorithm proceeds essentially by repetition of the labelling process described earlier. Each repetition takes place at line 4. What happens after each application of this labelling process depends upon whether or not the sink becomes labelled. If it becomes labelled then the edge flows along the resultant augmenting path are incremented in line 6, exactly as in the maximum-flow algorithm. If the required final value of the network flow, say V', is attained in this process then the algorithm terminates. If in the labelling process the sink does *not* become labelled, then two actions take place.

Fig. 4.8. Minimum-cost flow algorithm.

```
 1.  for all u ∈ V do π(u) ← 0
 2.  for all (u, v) ∈ E do f(u, v) ← 0
 3.  TEST ← true
 4.  Carry out the labelling process
 5.  if the sink is labelled then
         begin
 6.      Modify the edge flows, stopping if V = V'
 7.      TEST ← true
 8.      goto 4
         end
 9.  if the sink is not labelled then
         begin
10.      if TEST then stop if V saturates the network
11.      Modify the vertex numbers, π(u)
12.      TEST ← false
13.      goto 4
         end
```

First, if in the previous labelling process a flow incrementation resulted, then a check is made to see if the current flow is a maximum for the network. This is most easily achieved by noting whether or not the cut produced by the last labelling process is saturated. Of course, if a maximum flow for the network has been achieved then the algorithm terminates. The boolean variable *TEST*, used in line 10 and which is assigned to in lines 3, 7 and 12, is simply a device which, for complexity reasons, ensures that the test for saturation is only carried out if the previous labelling process caused a flow incrementation whilst the present labelling process does not. The other action that takes place if the sink does not become labelled is a modification of the vertex numbers in line 11. For the present we take this modification to be:

$$\pi(u) \leftarrow \pi(u) + 1$$

if, and only if, u is not labelled in the last application of the labelling process. Notice that this modification always increments $\pi(y)$ and that this implies, through (*a*), that p is incremented.

With the above description of the algorithm we can present the following theorem:

Theorem 4.8. The minimum-cost flow algorithm finds a minimum-cost flow.

Proof. Initial values of zero for the vertex numbers and for the edge flows ensure that at the outset, when $p = 0$, the complementary slackness conditions are satisfied. We now show that these conditions stay satisfied throughout the course of the algorithm.

The algorithm consists of a sequence of applications of the labelling procedure, each application is followed either by a flow change or by a vertex number change. Flow changes cannot affect the complementary slackness conditions because edge flows are only changed on edges (u, v) for which $\pi(u) - \pi(v) = a(u, v)$, whereas the slackness conditions (*b*) and (*c*) only apply to edges for which $\pi(u) - \pi(v) \neq a(u, v)$. Now consider vertex number changes. Primed quantities will denote values after an update of vertex numbers. For the complementary slackness conditions to hold we must show that:

$$\pi'(x) = 0, \quad \pi'(y) = p + 1 \qquad \text{(i)}$$

$$\pi'(v) - \pi'(u) < a(u, v) \Rightarrow f'(u, v) = 0 \qquad \text{(ii)}$$

$$\pi'(v) - \pi'(u) > a(u, v) \Rightarrow f'(u, v) = c(u, v) \qquad \text{(iii)}$$

In the application of the labelling process just prior to a vertex number update no augmenting path is found, so that, whilst the source is labelled, the sink is not. Thus (i) follows from the relabelling rule. We now prove (ii), the proof of (iii) being similar. If $\pi'(v)-\pi'(u) < a(u, v)$ then, since the vertex numbers are integer, it must be the case that $\pi(v)-\pi(u) \leqslant a(u, v)$. In the case of strict inequality (u, v) was not a candidate for flow change so that $f'(u, v) = f(u, v) = 0$ and (ii) holds. In the case of equality we have that $\pi'(v) = \pi(v)$ and that $\pi'(u) = \pi(u)+1$. Thus (u, v) has v labelled and u unlabelled. But since $\pi(v)-\pi(u) = a(u, v)$ it follows that $f'(u, v) = 0$, otherwise u would have been labelled from v. Hence (ii) again holds.

So the minimum-cost flow algorithm ensures that the complementary slackness conditions are satisfied throughout its operation. Therefore, for each value of $p = 0, 1, 2, \ldots$, the value of the expression

$$(pV - \sum_{(u, v)} a(u, v) f(u, v))$$

is maximised. As we observed earlier, each new addition to the flow has a unit cost of transportation from x to y equal to the latest value of p. Thus it is always the case that maximisation of the expression

$$(pV - \sum_{(u, v)} a(u, v) f(u, v))$$

is equivalent to minimising $\sum_{(u, v)} a(u, v) f(u, v)$. The algorithm terminates either when a desired overall flow is achieved in line 6, or when a flow is detected in line 10 which will be a maximum possible for the network. In the latter case we obtain a maximum flow at minimum cost. ∎

Let us now consider the complexity of the algorithm. As we remarked earlier, the algorithm can be viewed as a sequence of flow maximisations over subgraphs of the network (containing edges for which

$$(\pi(v)-\pi(u)) = a(u, v)$$

only), one for each of the values $p = 0, 1, 2, \ldots$. As we saw when describing the maximum-flow algorithm, each maximisation has an execution time which is polynomial in the size of the subgraph and, therefore, in the size of the network. Of course, many of the maximisations are likely to be trivial, contributing nothing to the overall network flow. In each such maximisation, the labelling process fails to label the sink even once for the current value of p. It is an easy matter to speed up the algorithm, causing it to by-pass many of these situations. This may be achieved (see exercise 4.9) by incrementing the vertex numbers by an amount which guarantees that at least one more vertex gets labelled in the next application of the labelling process. In this way there can never be more than n consecutive applications of the labelling process which result in no flow augmentation. Each flow augmentation adds at least one to the network

Fig. 4.9. An application of the minimum-cost flow algorithm.

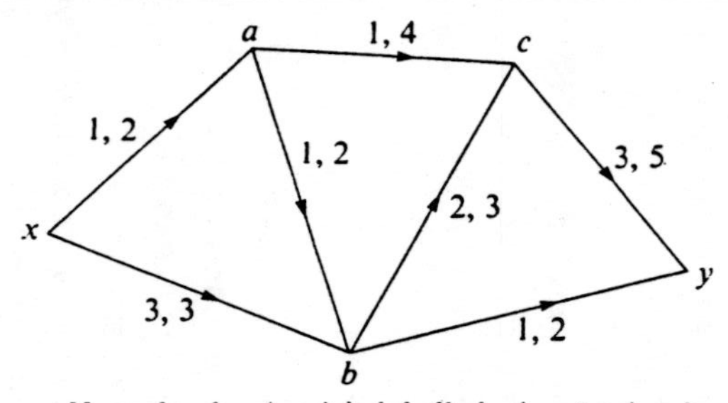

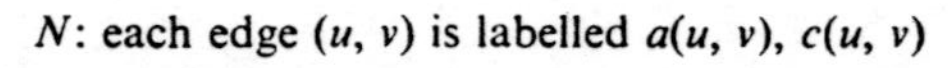

N: each edge (u, v) is labelled $a(u, v)$, $c(u, v)$

Iteration	$\pi(x)$	$\pi(a)$	$\pi(b)$	$\pi(c)$	$\pi(y) = P$	Edges effective in labelling	Labelled vertices	$f(x,a)$	$f(x,b)$	$f(a,c)$	$f(a,b)$	$f(b,c)$	$f(b,y)$	$f(c,y)$
0	0	0	0	0	0	none	x	0	0	0	0	0	0	0
1	0	1	1	1	1	(x, a)	x, a	0	0	0	0	0	0	0
2	0	1	2	2	2	$(x, a), (a, b), (a, c)$	x, a, b, c	0	0	0	0	0	0	0
3	0	1	2	2	3	$(x, a), (a, b), (a, c), (b, y)$	x, a, b, c, y	2	0	0	2	0	2	0
4	0	1	2	2	3	none	x	2	0	0	2	0	2	0
5	0	2	3	3	4	$(x, b), (b, a), (a, c)$	x, b, a, c	2	0	0	2	0	2	0
6	0	2	3	3	5	$(x, b), (b, a), (a, c)$	x, b, a, c	2	0	0	2	0	2	0
7	0	2	3	3	6	$(x, b), (b, a), (a, c), (c, y)$	x, b, a, c, y	2	2	2	0	0	2	2
8	0	2	3	3	6	(x, b)	x, b	2	2	2	0	0	2	2
9	0	3	3	4	7	(x, b)	x, b	2	2	2	0	0	2	2
10	0	4	3	5	8	$(x, b), (b, c), (c, y)$	x, b, c, y	2	3	2	0	1	2	3
11	0	4	3	5	8	none	x	Network saturated $F(N) = 5$						

flow. Thus there can never be more than V' flow augmenting maximisations interspersed between at most nV' maximisations which are not flow augmenting. Thus the complexity of the minimum-cost flow algorithm is polynomial in n, $|E|$ and the required network flow V'.

In figure 4.9 we can see an application of the minimum-cost flow algorithm. The table shows the vertex numbers and edge flows which follow successive applications of the labelling procedure. Finally, a maximum flow for the network is obtained. Of course, the flows obtained at each earlier stage of execution are minimum-cost flows for the current values of $F(N)$.

The minimum-cost flow algorithm, like several others which solve the same problem (see, for example, Busacker & Gower[7] and Iri[8]), works on the principle of incrementing the network flow along augmenting paths of minimum cost. This works because of an underlying theorem (proof of which is implicit in our verification of the present algorithm) that such an incrementation to a minimum-cost flow results also in a minimum-cost flow.

The algorithm described in this section is not as general as it might be. For example it cannot cope directly with non-zero minimum edge capacities, nor can it handle non-positive edge costs. These particular drawbacks can be avoided by employing another algorithm, also due to Ford & Fulkerson,[9] and which is known as the *out-of-Kilter* algorithm. For our purposes, we shall be satisfied with the minimum-cost flow algorithm described here.

4.5 Summary and references

Amongst others, the problems of maximising the flow and finding minimum-cost flows in networks, for which algorithms were described in this chapter, have long been of interest in operations research. The question of finding efficient algorithms has more recently been the interest of computer scientists as the references that follow make clear. Networks have been the subject of many variations and generalisations as their applications have warranted. Two variations not mentioned in the chapter or hinted at in the exercise section are as follows. The first is that *traversal times* may be associated with the edges; one problem might then be to send all the flow units across the network within a specified time. The second is to regard each edge as an *amplifier* so that the volume of flow is enhanced (or perhaps diminished) according to some parameter as any edge is traversed; one problem might then be to maximise the flow arriving at the sink for a given flow leaving the source. Chapter 4 of Minieka[12] provides an introduction to these generalisations. Chapter 4 of Lawler[13] and chapters 5 and 6 of Even[14] also provide general background for the material of this chapter. A third important generalisation (Lawler, for example, provides an introduction) concerns *multicommodity flows*. Several types of goods simultaneously traverse the network using the same edges but leave and arrive at their own sources and sinks. From the analysis point of view this creates much greater difficulty than is the case for single commodity flow and much work still needs to be done. For example, there is no result like the max-flow, min-cut theorem for general networks given multicommodity flow.

As was implied in the introductory paragraphs of this chapter, flow techniques have interesting applications in combinatorial problems. Apart from the instances mentioned there, others may be found amongst the exercises that follow. In dealing with the particular problem of connectivity in section 4.3 we drew upon Dantzig & Fulkerson.[5]

[1] Ford, L. R., Jr & Fulkerson, D. R. 'Maximal flow through a network', *Canad. J. Math.* **8**, 399–404 (1956).

[2] Edmonds, J. & Karp, R. M. 'Theoretical improvements in algorithmic efficiency for network flow problems', *JACM*, **19**, 248–62 (1972).

[3] Karzanov, A. V. 'Determining the maximal flow in a network by the method of preflows', *Soviet Math. Dokl.*, **15**, 434–7 (1974).

[4] Malhotra, V. M., Pramodh Kumar, M. & Maheshwari, S. N. 'An $O(|V|^3)$ algorithm for finding maximum flows in networks', *Computer Science Program*, Indian Institute of Technology, Kanpur 208016, India (1978).

[5] Dantzig, G. B. & Fulkerson, D. R. 'On the max-flow, min-cut theorem of networks', *Linear Inequalities and Related Systems*. Annals of Math. Study 38. Princeton University Press, 215–21 (1956).

[6] Ford, L. R. & Fulkerson, D. R. *Flows in Networks*, Princeton University Press, 113 (1962).

[7] Busacker, R. G. & Gowan, P. J. 'A procedure for determining a family of minimal cost network flow patterns', *ORO Technical Paper*, **15** (1961).

[8] Iri, M. 'A new method of solving transportation network problems', *J. Op. Res. Soc. Japan*, **3**, 27–87 (1960).

[9] Ford, L. R. & Fulkerson, D. R. *Flows in Networks*, Princeton University Press, 162 (1962).

[10] Hadlock, F. O. 'Finding a maximum cut of a planar graph in polynomial time', *SIAM J. Comput.*, **4**, 221–5 (1975).

[11] Karp, R. M. 'Reducibility among combinational problems'. In: *Complexity of Computer Computations*, Plenum Press (eds: Miller & Thatcher), NY, pp. 85–103 (1972).

[12] Minieka, E. *Optimisation Algorithms for Networks and Graphs.* Marcel Dekker, NY (1978).

[13] Lawler, E. *Combinatorial Optimisation: Networks and Matroids.* Holt, Rinehart and Winston (1976).

[14] Even, S. *Graph Algorithms*, Computer Science Press (1979).

EXERCISES

4.1. A generalisation of (transport) networks as defined in section 4.1 is to have several sources and several sinks. Let $\{x_1, x_2, \ldots\}$ and $\{y_1, y_2, \ldots\}$, respectively, be the set of sources and the set of sinks of such a generalised network. If we wish to maximise the overall flow from $\{x_1, x_2, \ldots\}$ to $\{y_1, y_2, \ldots\}$ then we can still use the maximum-flow algorithm to do this. Before applying the algorithm, however, we need to modify the network. This is done by adding a new source X and a new sink Y. Edges (X, x_1), (X, x_2), ... are then added from X to each original source and the edges (y_1, Y), (y_2, Y), ... are added from each original sink to Y. Each additional edge is given an infinite capacity.

Briefly justify the following statement. If the maximum-flow algorithm is used to maximise the flow from X to Y, then the flow obtained is also a maximum for the original network.

4.2. A manufacturer has two despatch points D_1 and D_2 for his goods which he sends to three market points M_1, M_2 and M_3 across the network shown below. Each edge is labelled according to the maximum flow of goods it can sustain. There is a market demand $\phi(M)$ at each of the market points as follows:

$$\phi(M_1) = 10, \quad \phi(M_2) = 8, \quad \phi(M_3) = 8$$

Can the network meet the demand? If a factory is sited at D_1 and another at D_2, determine (non-unique) separate outputs in order to meet the situation.

(Use the construction of exercise 4.1 and maximise the network flow.)

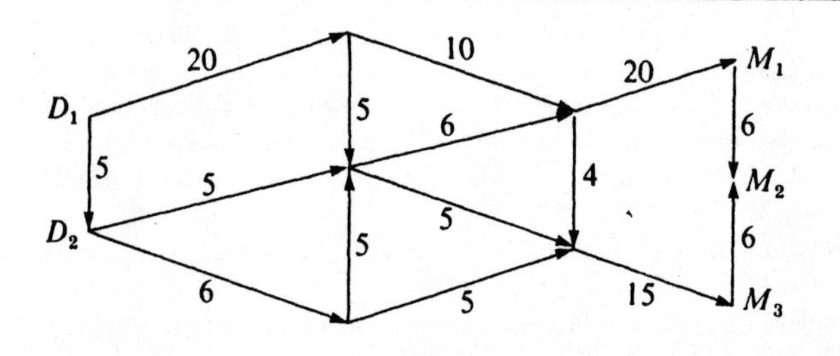

4.3. (*a*) We wish to construct, if it exists, a directed simple (not necessarily connected) graph with n vertices $\{v_i\}$, where each vertex v_i has a specified out-degree $d^+(v_i)$ and a specified in-degree $d^-(v_i)$. Show that the following procedure achieves this objective.

First construct a network N with the vertex-set $\{X, Y, a_1, a_2, \ldots, a_n, b_1, b_2, \ldots, b_n\}$ and edges

(X, a_i) for all i, where $c(X, a_i) = d^+(v_i)$
(b_i, Y) for all i, where $c(b_i, Y) = d^-(v_i)$
(a_i, b_j) for all $i \neq j$, where $c(a_i, b_j) = 1$

Now maximise the flow from X to Y. If the maximum flow saturates all edges from X and all edges to Y then a digraph satisfying the requirements exists. It is obtained from N by removing X and Y, and each v_i is formed by coalescing a_i with b_i.

(*b*) A question related to (*a*) concerns the *excursion problem.* R families go on an excursion in S vehicles. There are f_i people in the ith family and v_j seats in the jth vehicle. Is it possible to arrange the seating so that no two members of the same family are in the same vehicle? Describe how this question may be answered.

4.4. A generalisation of (transport) networks as defined in section 4.1 is to associate a minimum edge capacity $\bar{c}(u, v)$ with each edge (u, v), so that a feasible flow must satisfy:

$$\bar{c}(u, v) \leqslant f(u, v) \leqslant c(u, v) \quad \text{for all} \quad (u, v) \in E$$

and

$$\sum_u f(u, v) = \sum_u f(v, u) \quad \text{for all} \quad v \in V, v \neq x, y$$

(*a*) If for all (u, v), $\bar{c}(u, v) = 0$ as in section 4.1, then some feasible flow exists. However, for the present generalisation a feasible flow need not exist. Construct a simple example to show this.

(*b*) Suppose that a feasible flow F does exist for the network N. Starting from F show, as was done for the proof of theorem 4.3, that we can augment to a maximum value given by:

$$\max F(N) = \min(K(P, \bar{P}) - \sum_{\substack{u \in \bar{P} \\ v \in P}} c(u, v)).$$

(*c*) Show that flow F described in (*b*) can be reduced to a minimum value given by:

$$\min F(N) = \max\Big(\sum_{\substack{u \in P \\ v \in \bar{P}}} \bar{c}(u, v) - \sum_{\substack{u \in \bar{P} \\ v \in P}} c(u, v)\Big)$$

The reduction from F is achieved by 'increasing' the flow along a path from the *sink to the source.* Such paths may be found using a labelling process, similar to that described in theorem 4.2, which starts with a labelling of the sink.

4.5. One of the most popular uses of networks is in the scheduling of large, complicated tasks. Each edge of a *PERT* (Program Evaluation and Review Technique) digraph represents a subtask and its weight is the time required to complete that subtask. Each edge (u, v) represents a subtask which can only be started when those subtasks represented by edges incident *to u* have been completed. The digraph has one vertex with zero in-degree (the *start* vertex), and one vertex with zero out-degree (the *termination* vertex) and every vertex is on some path from the start vertex to the termination vertex.

(*a*) A properly constructed *PERT* digraph contains no directed circuits. Moreover (see (*b*)), it is useful to numerically label the vertices such that for every edge (u, v), $u < v$. This labelling is called a *topological sorting*. Verify that the following algorithm creates a topological sorting and checks that the digraph is acyclic, Initially every vertex of the digraph N is unlabelled.

1. $i \leftarrow 1$
2. look for a vertex $v \in N$ with zero in-degree
   if no such vertex exists **then goto** 5
   if such a vertex exists **then**
      begin
      label v with i
      redefine N: $N \leftarrow (N - v)$
      end
3. $i \leftarrow i + 1$
4. **goto** 2
5. **if** N contains no vertices
   then stop (the acyclic graph has a topological sorting)
   else stop (the graph contains a directed circuit)

(*b*) An important parameter in any *PERT* digraph is the length of the longest path from the start to the termination vertex. Such a path is called a *critical path*, and its length represents the shortest time within which the overall task can be completed. For this reason the analysis is sometimes called *CPM* (Critical Path Method). Verify that, if the vertices of N are topologically sorted, then the following algorithm finds a critical path length. In fact it finds the longest path length $L(i)$ from the start vertex to each vertex i of the network. Task (j, i) takes a time $w(j, i)$ to complete.

1. **for** $i = 1$ **to** n **do** $L(i) \leftarrow 0$
2. **for** $i = 2$ **to** n **do**
   for all j such that $(j, i) \in E$ **do**
      $L(i) \leftarrow \max\,(L(i), (L(j) + w(j, i))$

4.6. One version of the well-known *knapsack problem* is as follows. There are N items, the jth is denoted by I_j and it has an integer *weight* w_j and an integer *value* v_j. The problem is to place a number of these items into a knapsack which can take a maximum total weight of W and to do this so that the items in the knapsack have a maximum combined value. This problem can be transformed into one of finding a maximum-length path in a weighted acyclic network N (see the previous exercise) by the following construction.

N contains $N.(W+1)$ vertices each denoted by $v(i, j)$. The range of i is from 1 to N while j ranges from 0 to W. Every vertex $v(i, j)$ has two edges incident to it, one from $v((i-1), j)$ with zero weight and one from $v((i-1), (j - w_i))$ with weight v_i, provided these vertices exist. We now add a source X and the edges $(X, v(1, 0))$ with zero weight and $(X, v(1, w_1))$ with weight v_1. Finally we add a sink Y and the edges $(v(N, j), Y)$ for $j = 0, 1, \ldots, W$, all with zero weight.

Prove that the set of paths from X to $v(i, j)$ represents all possible subsets of $\{I_1, I_2, ..., I_i\}$ which have a total weight of j. Notice that one of the two edges incident to $v(i, j)$ represents the *absence* of I_i and the other represents the presence of I_i in a subset. Notice also that each path length is the combined value of the represented subset. It follows that the longest path from X to Y solves this knapsack problem.

4.7. Given the definition of $p_v(v_i, v_j)$ in section 4.3 show that if min $p_v(v_i, v_j)$ occurs for two adjacent vertices u and v in an incomplete graph G, then it also occurs for two non-adjacent vertices.
(Define $G' = G-(u, v)$ and let $p'_v(u, v)$ be the maximum number of vertex disjoint paths between u and v in G', then:
$p_v(u, v) = p'_v(u, v)+1$
Since $(u, v) \notin G'$ there exists, by theorem 4.6, a vertex-cut VC separating u from v in G', which has cardinality $C = p'_v(u, v)$. Now $C < (n-2)$ because if $C = n-2$ then $p_v(u, v) = n-1$. This is not possible because G is incomplete: for any two non-adjacent vertices a and b it must be that $p_v(a, b) \leqslant n-2$ and $p_v(u, v)$ is supposed to be a minimum over all pairs of vertices. It follows that there exists a vertex $w \neq u$ or v and which is not a member of VC. Suppose, without loss of generality, that VC separates w from u (as opposed to w from v) in G'. The result then follows by observing that $(VC \cup \{v\})$ must separate u from w in G.)

4.8. An employer wishes to hire, at a minimum retraining cost, n employees for different skilled work. He may choose from m candidates where $m \geqslant n$. The cost of retraining candidate C_i for job a_j is b_{ij}. Briefly justify the following minimum-cost flow formulation of the problem.
Construct a network N with the vertex-set $\{X, Y, c_1, c_2, ..., c_m, a_1, a_2, ..., a_n\}$ and the edges:
(X, C_i) for $i = 1, 2, ..., m$, where $c(X, C_i) = 1$, $a(X, C_i) = 0$
(a_j, Y) for $j = 1, 2, ..., n$, where $c(a_j, Y) = 1$, $a(a_j, Y) = 0$
(C_i, a_j) for all i, j where $c(C_i, a_j) = 1$, $a(C_i, a_j) = b_{ij}$
where for any edge (u, v), $c(u, v)$ is the capacity and $a(u, v)$ is the cost. Produce a maximum flow at minimum cost from X to Y. Employ each candidate C_i for which the flow $f(X, C_i) = 1$, and assign him for retraining to job a_j if $f(C_i, a_j) = 1$.

4.9. Suppose that in the minimum-cost flow algorithm the vertex numbers are not incremented by one but by Δ defined as follows:
$\Delta = \min(\delta_1, \delta_2) > 0$
where
$\delta_1 = \min\{(a(u, v)+\pi(u)-\pi(v)) | (u, v) \in E_1\}$
$\delta_2 = \min\{(\pi(v)-\pi(u)-a(u, v)) | (u, v) \in E_2\}$
$E_1 = \{(u, v) | u \in P, v \in \bar{P}, \pi(v)-\pi(u) < a(u, v)\}$
$E_2 = \{(u, v) | u \in \bar{P}, v \in P, \pi(v)-\pi(u) > a(u, v)\}$

and $(P, \bar{P})$ is the vertex-cut induced by the last application of the labelling process. Show that this modification ensures that the labelling process will label one more vertex in the next application than in the last. Thus show that there could not be more than n consecutive applications of the labelling process which result in no flow augmentation.

4.10. The question: 'Does the network N contain a cut of capacity less than K?' can be answered in polynomial time by an application of the maximum-flow algorithm and because of the max-flow, min-cut theorem. Suppose that N is planar. Construct a polynomial time algorithm to answer the question: 'Does the network N contain a cut of capacity *greater than* K?'
(See Hadlock.[10] However, if N is non-planar the problem is *NP*-complete, see Karp.[11])

5

Matchings

A *matching* of a graph is any subset of its edges such that no two members of the subset are adjacent. Interest in matchings arises in a direct and natural way as described in some of the exercises at the end of this chapter. Also the search for certain matchings can be an important subtask for some larger problems such as the problem of the Chinese postman described in chapter 6. Central to the content of this chapter is the description of a maximum-cardinality matching algorithm in section 5.2, and the description of a maximum-weight matching algorithm in section 5.3.

5.1 Definitions

A *matching* of a graph $G = (V, E)$ is any subset of edges $M \subseteq E$ such that no two elements of M are adjacent. For example, some matchings of the graph of figure 5.1 are $\{e_1\}$, $\{e_1, e_5, e_{10}\}$, $\{e_2, e_7, e_{10}\}$ and $\{e_4, e_6, e_8\}$. Clearly, any subset of a matching is also a matching.

Fig. 5.1

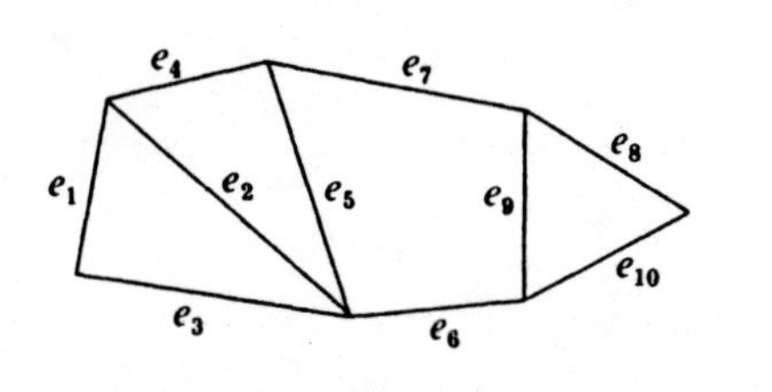

A *maximum-cardinality matching* is a matching which contains a maximum number of edges. A *perfect matching* is a matching in which every vertex of the graph is an end-point of some element of the matching. Not every graph contains a perfect matching. In section 5.2.1 we provide a necessary and sufficient condition for a graph to contain a perfect matching. Clearly, if a graph does contain a perfect matching M, then M will be a maximum-cardinality matching and any maximum-cardinality matching will be a perfect matching.

In a bipartite graph G with bipartition (V', V''), a complete matching of V' onto V'', is a matching M in which every element of V' is an end-point of an element of M. In exercise 5.3 we describe a sufficient and necessary condition for a bipartite graph to contain a complete matching. If a bipartite graph contains a complete matching M, then M is clearly a maximum-cardinality matching, and a maximum-cardinality matching is of course complete.

In a weighted graph, a *maximum-weight matching* is a matching for which the sum of the edge-weights is a maximum.

5.2 Maximum-cardinality matchings

Consider first the special case of bipartite graphs. For these we can use a simple method, utilising flow techniques (see chapter 4), to find a maximum-cardinality matching. Let $G = (V, E)$ be a bipartite graph with the bipartition (V_1, V_2). We construct a network G' from G as follows:

(1) Direct all edges from V_1 to V_2.
(2) Add a source x and a directed edge from x to each element of V_1.
(3) Add a sink y and a directed edge from each element of V_2 to y.
(4) Let each edge (u, v) have a capacity $c(u, v) = 1$.

Fig. 5.2

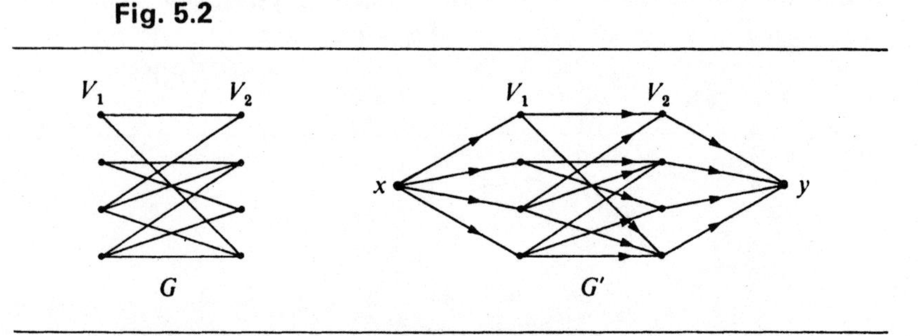

Given such a construction, which is illustrated in figure 5.2, we can find a maximum-cardinality matching M by maximising the flow from x to y. M then consists of those edges linking V_1 to V_2 which carry a flow of one unit. If some matching M' exists such that $|M'| > |M|$ then we could construct a flow of value $|M'|$, which is greater than the value of the flow found, by sending one unit of flow along each path $((x, u), (u, v), (v, y))$ for all $(u, v) \in M'$.

We now consider general graphs. In the case of bipartite graphs the algorithm we shall describe is just another view of the algorithm just outlined. We need the following definitions. If $M \subseteq E$ is a matching for $G = (V, E)$, then any vertex $v \in V$ is called a *free* vertex if it is *not* an end-point of any element of M. An *alternating path* is a simple path in G whose edges alternately belong to M and to $(E-M)$. An *augmenting path* with respect to M is an alternating path between two free vertices. Denoting the cardinality of M by $|M|$, notice that if G contains an augmenting path P, then a matching M' can be found such that

$$|M'| = |M|+1$$

simply by reversing the rôles of those edges in P. Those edges not in M are placed in M, whilst those in M are removed from M. Thus in figure 5.1, if $M = \{e_3, e_8\}$, then an augmenting path can be traced along the sequence of edges (e_1, e_3, e_5). Reversing edge rôles we obtain $M' = \{e_1, e_5, e_8\}$. The following theorem underlines the importance of augmenting paths with respect to maximum-cardinality matchings.

Theorem 5.1. There is an M-augmenting path if and only if M is not a maximum-cardinality matching.

Proof. Clearly, if $G = (V, E)$ contains an M-augmenting path then we can construct M' such that $|M'| = |M|+1$. Hence M cannot be a maximum-cardinality matching.

Conversely, suppose that M is not a maximum-cardinality matching. We shall show that G contains an M-augmenting path. Let M' be a maximum-cardinality matching so that $|M'| > |M|$. Also let

$$G' = (V, M \oplus M')$$

In other words, G' contains those edges which are either in M or in M' but not in both. Notice that:

(*a*) G' has more edges from M' than from M,

and

(*b*) each vertex of G' is incident to at most one edge from M' and at most one edge from M.

It follows that each component of G' is either an isolated vertex or an alternating path. Moreover, at least one component (because of (*a*)) must have more edges from M' than from M. Such a component is an M-augmenting path. ■

The last theorem naturally suggests an algorithm to find a maximum-cardinality matching. Starting from an arbitrary matching (the null matching for want of another) we repeatedly carry out augmentations along M-augmenting paths until no such path exists. This process is bound to terminate because a maximum matching has finite cardinality and each augmentation increases the cardinality of the current matching by one. The only practical problem is to specify a systematic search for M-augmentations. Before we do this it is interesting to emphasise the relationship between this algorithm applied to bipartite graphs and the flow method described earlier.

Consider the flow F obtained at some intermediate stage of maximisation in the flow method to find a maximum-cardinality matching in a bipartite graph. This flow corresponds to a matching M in G. M consists of those edges with a non-zero flow. Any flow augmenting path with respect to F coincides with some M-augmenting path in G. Moreover, if we separately carry out a flow augmentation in the flow algorithm and an M-augmentation in the second algorithm (with respect to corresponding augmentation paths), then there is still a one-to-one coincidence between flow- and M-augmenting paths. Moreover, a flow augmenting path exists if and only if an M-augmenting path does. Clearly, both algorithms are different aspects of essentially the same process. The flow point of view as described here cannot be used for non-bipartite graphs although the idea of M-augmentations has quite general applicability. Incidentally, Edmonds[1] has shown that non-bipartite matching can be handled through so-called *bidirected network flow theory*. We do not, however, detail that approach here.

We return now to maximum-cardinality matching in general graphs broadly described earlier. The crux of the algorithm is the means by which augmenting paths are found. The following procedure by which such paths may be found is inspired by Edmonds.[2] The procedure constructs a search tree T rooted at some free vertex v. Any path in T starting at v is an alternating path in which the vertices are alternately labelled *outer* and *inner*. The root v is labelled *outer*. The procedure, which we call the M-augmenting path search procedure, *MAPS*(G), is shown in figure 5.3. T is externally initialised to be v, at which time v is labelled *outer*. There are, in fact, three possible exits from the procedure. These are to labels A, B and H. Only the exit to A indicates that an augmenting path has been found. In other words, that some leaf of T is found to be a free vertex. Let us describe each exit in turn.

Notice that *MAPS*(G) constructs a tree unless y is found to be labelled *outer* in which case an odd-length circuit has been found and this causes a jump to B. If y is *inner*, an even-length circuit has been detected. In this case (x, y) is not added to T and the procedure seeks to extend the tree from

Fig. 5.3. The M-augmenting path search procedure $MAPS(G)$.

1. Choose an *outer* vertex $x \in T$ and some edge (x, y) not previously explored. Deem (x, y) to have been explored. If no such edge exists **goto** H.
2. If y is *free* and unlabelled, add (x, y) to T. **goto** A.
3. If y is *outer* add (x, y) to T. **goto** B.
4. If y is *inner* **goto** 1
5. Let (y, z) be the edge in M with the end-point y. Add (x, y) and (y, z) to T. Label y *inner* and z *outer*. **goto** 1.

some other *outer* vertex. Why does the procedure terminate on detecting an odd-length circuit? Consider the graph G of figure 5.4(a) in which the edge labelled M constitutes a matching. Clearly, G contains an M-augmenting path, namely, $P = (v_1, v_2, v_3, v_4)$. If we now call $MAPS(G)$ with T initialised to v_1 and if line 2 is first executed with $y = v_3$, then the procedure terminates with a jump to B. T will then be as shown in figure 5.4(b). Now P cannot be found because v_2 is labelled *inner*. The presence of odd-length cycles introduces ambiguities in alternating path searches.

Fig. 5.4

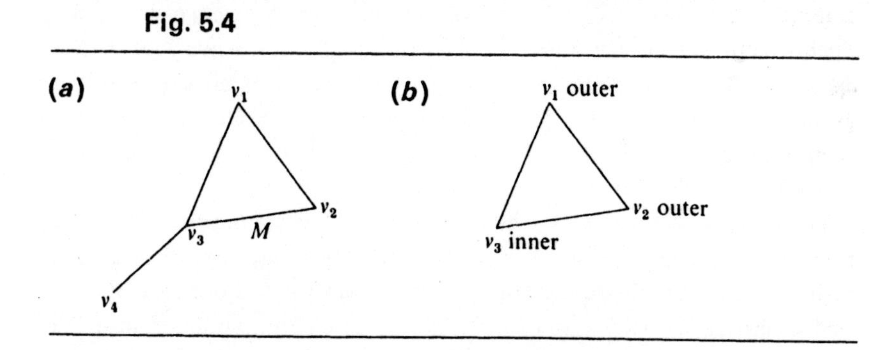

This is because any vertex v_j on such a circuit may be labelled either *outer* or *inner* depending upon which direction around the circuit v_j is approached from when tracing a path in T from the root to v_j. If v_j is labelled *inner*, then T cannot be extended from v_j and possible augmenting paths can go undetected. In fact the maximum-cardinality matching algorithm which uses $MAPS(G)$ keeps all options open regarding any odd-length cycle C. A new graph is constructed by shrinking C to form a single vertex which is labelled *outer*. The algorithm then continues with another call of the procedure $MAPS(G)$. All previous labels of *outer* and *inner*, except those on C, are carried forward. If subsequently a leaf of T is found that is a free vertex, then the augmenting path found might pass through one or more of these artificially created *pseudo-vertices.* Indeed, notice that an odd-length circuit may itself contain pseudo-vertices and so on. We return to this when describing the exit to label A. Edmonds gave the name *blossom* to an odd-length circuit found by the procedure. A bipartite graph contains only even-length circuits. Without the need to handle blossoms, the algorithm could be considerably simplified for bipartite graphs.

Now consider the exit to H. In this situation T cannot be extended. Each alternating path traced from the root of T is stopped at some *outer* vertex. The only free vertex is the root of T. T is now called a *Hungarian tree.* Some vertices labelled *outer* may be pseudo-vertices, but each vertex labelled *inner* is an ordinary vertex. It is crucial to notice that edges connecting vertices in T to vertices *not* in T can only be attached to *inner* vertices of T. Otherwise some *outer* vertex must be connected to a free vertex or T must be extendable from the vertex. We can see that no vertex in a Hungarian tree can possibly occur in an augmenting path. T only contains one free vertex, so that if some vertex of T is on an augmenting path, then this path must enter T along an edge (not in M) at an *inner* vertex. Thereafter the path must alternately visit *outer* and *inner* vertices, entering the former along edges of M and the latter along edges not in M. Such a path can neither reach the root of T nor can it leave T. Thus on exiting to H, the maximum-cardinality algorithm can remove T from the graph in the current search for an M-augmenting path. Of course, if M is eventually augmented then T must be restored to G before the next augmenting path is sought.

Finally consider the exit to A. T contains an alternating path from the root of T to some other free vertex. However, this may pass through one or more shrunken blossoms. These can be expanded and one side of each odd-length circuit (the side of even-length) can be interpolated into the path. This continues until no blossoms exist on the augmenting path. Of course, each expansion may expose other pseudo-vertices which were created earlier than the one just expanded. However, eventually no blossoms remain and an augmentation of M becomes possible.

In describing the exits from *MAPS* we have more or less described the maximum-cardinality matching algorithms. Before describing an application of this algorithm we have some comments to make. The algorithm is outlined in figure 5.5. The bulk of the algorithm consists of the **while** statement, lines 3–17 inclusive, which iterates once for each augmentation of the matching. Line 1 initialises M to be the null matching. In line 6 a stack is initialised. This stack is used to store *blossoms* and *Hungarian trees* in the order in which they are found through lines 10 and 11. In this way the graph can be properly reconstructed within line 15. Each iteration of the while statement starts by determining the free vertices with respect to the current matching M in line 4. Then the **for** statement commencing at line 5 attempts to find an augmenting path utilising the procedure *MAPS*. The labels B, H and A denote the exits from *MAPS*, as described earlier. Line 12 is only reached if either no free vertices exist with respect to the

Fig. 5.5. A maximum-cardinality matching algorithm.

---

1. $M \leftarrow \emptyset$
2. an augmenting path exists ← **true**
3. **while** an augmenting path exists **do**
    begin
4.   determine the free vertices $\{v_i\}$ with respect to M
5.   **for each** v_i **do**
        begin
6.   empty the stack
7.   deem each edge of G to be unexplored
8. $T \leftarrow v_i$, label v_i *outer*

L: 9.   *MAPS*(G)

B: 10.   Place the *Blossom* found on the stack, shrink it in G and label the resultant vertex *outer* and if it contains the root label it *free*. **goto** L.

H: 11.   Place the *Hungarian tree* found on the stack and remove it from G.

        end

12.   Output M.
13.   an augmenting path exists ← **false**
14.   **goto** S

A: 15.   Identify the M-augmenting path $P \in T$. Empty the stack, expanding G with each popped item. If the item is a blossom corresponding to a pseudo-vertex on P, interpolate into P the appropriate even-length section of the blossom.

16.   Augment M by interchanging the edge rôles in P

S: 17.   **end**

---

current matching, or if no augmenting path is found from any of the existing free vertices. The matching then has maximum cardinality and the algorithm terminates.

Figure 5.6 illustrates an example application of the maximum-cardinality matching algorithm. For G shown there, we detail the three iterations required to maximise the matching. The vertices of G are identified by the numerals 1–5. In choosing a next edge (u, v) to explore from u in the construction of each T, first u and then v are conveniently chosen to be the first (numerically) available vertex. The first iteration naturally discovers that the first edge explored is a path between free vertices. Augmentation adds this edge to M. The second iteration discovers, with the addition of edge (2, 3) that T contains a blossom with vertices 1, 2 and 3. This is shrunk to form a pseudo-vertex, 6. Exploration of the edge from 6 to 4 discovers that 4 is a free vertex. Hence an augmenting path P has been found. Expansion of 6 interpolates the even-length portion ((3, 1), (1, 2)) of the blossom into P. Augmentation gives

$$M = ((3, 1), (2, 4))$$

Of course, this must be a maximum matching because only vertex 5 is free. However, the algorithm, as described in figure 5.5, only terminates when it fails to find an augmenting path. This is what happens in the final iteration. In generating T from the root, vertex 5, a pseudo-vertex 7 containing vertices 3, 2 and 4 is created. Then finally, with the addition of the edge (5, 7), T becomes a single pseudo-vertex 8 containing vertices 1, 5 and 7. At the same time the contractions in G, associated with the shrinking of blossoms, have reduced G to the single vertex 8. Thus T must now be a Hungarian tree (albeit a single vertex) and the algorithm terminates.

The maximum-cardinality matching algorithm, as is easily seen, is a polynomial time algorithm. Its complexity is dominated by the accumulated costs of finding augmenting paths. There can be no more than $O(n)$ augmentations. To find an augmenting path at most $O(n)$ free vertices have to be considered. In considering each free vertex we construct a search tree T. This construction, including handling blossoms, requires no more than $O(|E|)$ steps. Thus even a cursory inspection shows that at most $O(n^2|E|)$-time is required. In fact we can easily improve this estimate to $O(n|E|)$. Notice that if a free vertex fails to yield an augmentation path, then a Hungarian tree is found. As we have seen, no edge incident to a vertex of such a tree can be on an augmentation path. As indicated in line 11 of the algorithm, we remove these edges from G as the search for an augmentation path proceeds to the next free vertex. Thus only $O(|E|)$-time is required to find an augmentation path. Micali & Vazirani[3] have described how maximum-cardinality matchings may be found in $O(\sqrt{n}|E|)$-time.

Verification of Edmond's maximum-cardinality matching algorithm is embedded in our description of it. It works by successively reducing the number of M-augmenting paths. Such a path, if it exists, will be found through the *MAPS* procedure. Eventually, no further paths can be found and by theorem 5.1 the final matching must be of maximum cardinality.

Fig. 5.6. An application of the maximum-cardinality matching algorithm.

1 3 2 5 4 *G*

First iteration

Free vertices: 1 2 3 4 5

$M = \emptyset$

T: 1 (root) *outer* — 2 (free)

$P = (1, 2)$ Augmentation gives $M = ((1, 2))$

Second iteration

Free vertices: 3 4 5

$M = ((1, 2))$

T: 3 (root) *outer*, 1 *inner*, 2 *outer* → 6 (1 2 3) *outer*, 4 (free)

$P = (\text{(1 2 3)}, 4) \rightarrow ((3, 1), (1, 2)\ (2, 4))$ Augmentation gives $M = ((3, 1), (2, 4))$

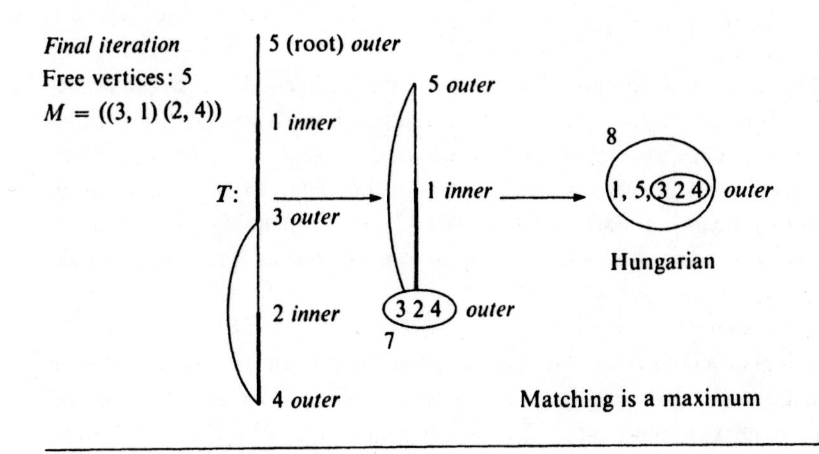

5.2.1. Perfect matchings

If every vertex of a graph is the end-point of an edge in a matching M, then M is said to be a *perfect* matching. Clearly, if a perfect matching exists for some graph G, then the result of applying the maximum-cardinality matching algorithm of the previous section to G will be to find a perfect matching. Obviously, a necessary but not sufficient condition for a graph G to have a perfect matching is that G has an even number of vertices. The following theorem is due to Tutte,[4] although our proof follows Lovász.[5] It provides a necessary *and* a sufficient condition for G to have a perfect matching. Within the theorem $\Phi(G-V')$ denotes the number of components of the graph $(G-V')$ containing an odd number of vertices. Such a component is called an *odd* component and, naturally, if a component of $(G-V')$ contains an even number of vertices then it is an *even* component. V' is any subset of V where $G = (V, E)$.

Theorem 5.2. $G = (V, E)$ has a perfect matching if and only if:

$$\Phi(G-V') \leqslant |V'| \quad \text{for all} \quad V' \subset V$$

Proof. Let us first suppose that G has a perfect matching M. We denote by G_i the ith odd component of $(G-V')$ where $i = 1, 2, \ldots, k$. Because G_i contains an odd number of vertices, some vertex of G_i (say v_i) must be matched by an edge of M to some vertex (say u_i) of V'. Now, because $\{u_1, u_2, \ldots, u_k\} \subseteq V'$, it follows that:

$$\Phi(G-V') = k = |\{u_1, u_2, \ldots, u_k\}| \leqslant |V'|$$

To complete the proof we need to show that if G satisfies

$$\Phi(G-V') \leqslant |V'|$$

for all V' then G contains a perfect matching. We shall show that the supposition that G contains *no* perfect matching leads to a contradiction. By adding edges to G we construct a maximal graph G^* with no perfect matching. That is the addition of *any* further edge to G^* will make a perfect matching possible. Now $(G-V')$ is a spanning subgraph of (G^*-V') so that $\Phi(G-V') \geqslant \Phi(G^*-V')$. It follows that G^* satisfies $\Phi(G^*-V') \leqslant |V'|$ for all V'.

Let us denote by V_0 the set of vertices of degree $(|V|-1)$ in G^*. If $V_0 = V$ then G^* is complete and so, contrary to our assumption, has a perfect matching. Notice in this respect that G^* has an even number of vertices, because if we let V' be the empty set in $\Phi(G^*-V') \leqslant |V'|$ then $\Phi(G^*) = 0$. We assume from now on then that $V_0 \neq V$.

We now show that each component of (G^*-V_0) is a complete graph. Let G_i^* be such a component which we initially presume is *not* complete. It is a simple matter to show that any incomplete graph has three vertices v_1, v_2 and v_3 such that (v_1, v_2) and (v_2, v_3) are in the edge-set but that (v_1, v_3) is not. Also, because $v_2 \notin V_0$, there exists a vertex v_4 in $(G^* - V_0)$ such that (v_2, v_4) is *not* in the edge-set of G^*. This situation is illustrated in figure 5.7(*a*) where absent edges are indicated by dashed lines. G^* is maximal so that $(G^* + (v_1, v_3))$ has a perfect matching, say M_1, and $(G^* + (v_2, v_4))$ has a perfect matching, say M_2. Consider the component H_1 of the subgraph

Fig. 5.7

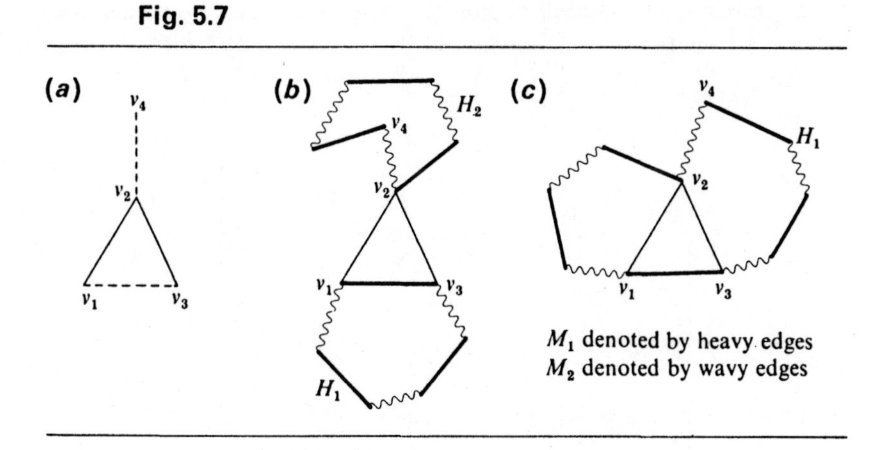

M_1 denoted by heavy edges
M_2 denoted by wavy edges

$(M_1 \oplus M_2)$ of $(G^* \cup \{(v_1, v_3), (v_2, v_4)\})$ which contains (v_1, v_3). Now $(v_1, v_3) \in M_1$ but $(v_1, v_3) \notin M_2$, so that H_1 consists of an even-length circuit of edges alternately in M_1 and in M_2. If H_1 does not contain (v_2, v_4), which is in M_2 but not in M_1, then (v_2, v_4) belongs to a different alternating cycle H_2. Clearly, H_1 and H_2 would be edge disjoint. This situation is shown in figure 5.7(*b*). The situation that both (v_1, v_3) and (v_2, v_4) are in H_1 is shown in figure 5.7(*c*).

We first consider the case of figure 5.7(*b*). G^*, which does not contain (v_1, v_3) or (v_2, v_4), clearly has a perfect matching containing the edges M_1 of H_1 and the edges M_2 of H_2. This contradicts the definition of G^*. Now consider the case of figure 5.7(*c*). Because of the symmetry of v_1 and v_3 we can assume that v_1, v_2, v_4 and v_3 occur in that order along the circuit H_1, as shown in the diagram. Again, G^* has a perfect matching which contains the edges of M_1 in the section $v_2, v_4, \ldots, v_3$ of H_1, the edge (v_2, v_3) and the edges of M_2 not in the section $v_2, v_4, \ldots, v_3$ of H_1. So again we have a contradiction of the definition of G^*.

Both cases provide contradictions so that it must be the case that each component of $(G^* - V_0)$ is a complete graph. Now $\Phi(G^* - V_0) \leqslant |V_0|$ so that $(G^* - V_0)$ has at most $|V_0|$ odd components. We can again construct a perfect matching in this case for G^* as follows. Each vertex of an even component of (G^*-V_0) can be matched by an edge of the component because the component is complete. Each vertex of an odd component of (G^*-V_0), except one, can be matched by an edge of the component because it is complete. The remaining vertex is then matched to a vertex in V_0. We then just need to match the remaining vertices of V_0. Now G^* has an even number of vertices, so that the remaining unmatched vertices of V_0 are even in number. The vertices can therefore be matched by edges of the complete subgraph induced by V_0. The perfect matching in this case is shown schematically in figure 5.8.

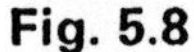

Fig. 5.8

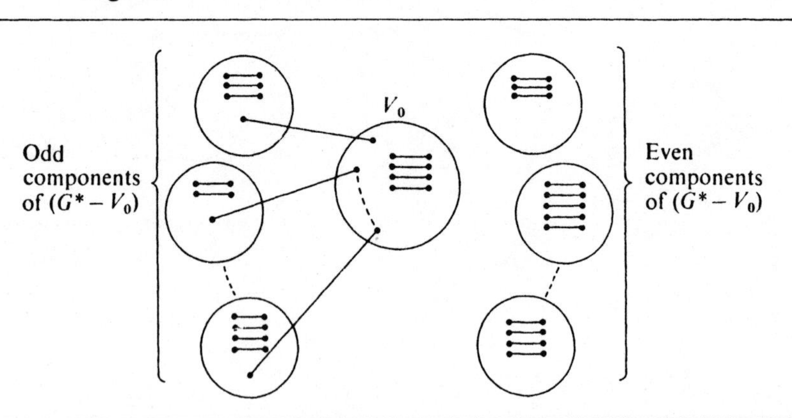

Thus contrary to our assumption G^* has a perfect matching and hence so does G. ■

The necessary and sufficient condition provided by theorem 5.2 does not lead to an efficient algorithm to determine whether or not a graph contains a perfect matching. This is because the number of subsets $V' \subset V$ is exponentially large. In contrast, the maximum-cardinality matching algorithm described earlier will of course answer the question 'does G contain a perfect matching' in polynomial time. For certain classes of highly symmetric graphs, theorem 5.2 can provide short proofs of the existence of perfect matching. See, for example, exercises 5.5 and 5.6.

5.3 Maximum-weight matchings

Just as is the case for maximum-cardinality matching, there are a number of specific and relatively simple algorithms to solve the maximum-weight matching problem in *bipartite* graphs. See, for example, Kuhn.[6] The algorithm we describe here which is due to Edmonds & Johnson[7] has a considerably simplified form for bipartite graphs. This simplification is left as an exercise for the reader.

Edmonds & Johnson's algorithm is a generalisation of the algorithm described in section 5.2 and our description proceeds along parallel lines.

A *weighted M-augmenting path* is an alternating path in which the sum of the weights of the edges not in the matching M is greater than the sum of the weights of the edges in M. Also, if either the first or the last edge of the path is not in M, then the edge has a free vertex as an end-point. Clearly, by interchanging the rôles of the edges of a weighted M-augmenting path (that is, those in M are removed from M and those not in M are placed in M), then a new matching of greater weight is obtained. Unlike the M-augmenting paths of the previous section, a *weighted M-augmenting path* may have edges in M as first and/or final edges. This induces the following definitions. A weighted M-augmenting path which contains:

(*a*) more edges not in M than in M is called a *strong augmenting path*,

(*b*) the same number of edges not in M as in M is called a *neutral augmenting path*,

(*c*) more edges in M than not in M is called a *weak augmenting path.*

Theorem 5.1 has the following counterpart:

Theorem 5.3. There is a weighted M-augmenting path if and only if M is not a maximum-weight matching.

Proof. Clearly, if a weighted M-augmenting path P exists in $G = (V, E)$ then we can interchange the rôles of the edges in P to obtain a matching of greater weight. Hence M cannot be a maximum-weight matching.

Conversely suppose that M is not a maximum-weight matching. We shall show that G contains a weighted M-augmenting path. Let M' be a maximum-weight matching, so that $w(M') > w(M)$. Let $G' = (V, M \oplus M')$. That is, G' contains those edges which are either in M or M' but not in both. Within G' each vertex is the end-point of at most one edge from M and at most one edge from M'. Thus each component of G' is a path (perhaps an even-length circuit) of edges alternately from M and M'. Since $w(M') > w(M)$ there must be at least one component of G' in which the sum of the edge-weights for edges in M' exceeds the sum of the edge-weights for edges in M. Such a component will be a weighted M-augmenting path. ■

Before describing Edmonds & Johnson's maximum-weight matching algorithm, we need to describe a linear programming (see the appendix) formulation of the problem. In the following formulation $x(u, v) = 1$ if $(u, v) \in M$ otherwise $x(u, v) = 0$.

$$\text{Maximise} \sum_{(u,v)} x(u, v)\, w(u, v)$$

subject to the constraints

$$\sum_{v} x(u, v) \leqslant 1 \quad \text{for all} \quad u \in V$$

$$\sum_{(u,v)\in R_k} x(u, v) \leqslant r_k \quad \text{for} \quad 1 \leqslant k \leqslant z$$

and the non-negativity conditions:

$$x(u, v) \geqslant 0 \quad \text{for all} \quad (u, v)$$

Each edge has a weight $w(u, v)$, so that the objective function is simply the weight of the matching. The first set of constraints states that no more than one edge in M is incident with any vertex u. In the second set of constraints R_k denotes the subgraph induced by any set of $(2r_k+1)$ vertices. We denote by z the number of these subgraphs. Clearly, there can be no more than r_k edges of R_k in M. As we shall demonstrate, this particular formulation provides a set of complementary slackness conditions which can be satisfied by an assignment of zero or one to each value $x(u, v)$.

The dual linear programming problem is expressed by the following. The dual variables y_v and z_k are, respectively, associated with the primal constraints for the vertex v and the subgraph R_k.

$$\text{Minimise} \sum_{v} y_v + \sum_{k} r_k z_k$$

subject to the constraints

$$y_u + y_v + \sum_{k:\,(u,v)\in R_k} z_k \geqslant w(u, v) \quad \text{for all} \quad (u, v)$$

and the non-negativity conditions:

$$y_v \geqslant 0 \quad \text{for all} \quad v$$

$$z_k \geqslant 0 \quad \text{for} \quad 1 \leqslant k \leqslant z$$

Notice that within the constraint for (u, v), the summation is over all k such that R_k contains (u, v).

The following complementary slackness conditions are provided by the primal–dual pair:

$$x(u, v) > 0 \Rightarrow y_u + y_v + \sum_{k:\,(u,v)\in R_k} z_k = w(u, v) \quad \text{for all} \quad (u, v)$$

$$y(u) > 0 \Rightarrow \sum_{v} x(u, v) = 1, \quad \text{for all} \quad u \in V$$

$$z_k > 0 \Rightarrow \sum_{(u,v)\in R_k} x(u, v) = r_k \quad \text{for all} \quad 1 \leqslant k \leqslant z$$

We shall refer to these three sets of conditions as respectively the *edge*, *vertex* and *odd-cardinality subset slackness conditions*.

The algorithm we describe starts with the null matching (that is, $x(u, v) = 0$ for all $(u, v) \in E$) and with dual variables:

$$y_s = W = \tfrac{1}{2} \max_{(u,v)} (w(u, v)), \quad \text{for all} \quad s \in V$$

$$z_k = 0, \quad \text{for } 1 \leqslant k \leqslant z$$

Thus at the outset, the constraints and non-negativity conditions for both primal and dual problems are satisfied and, in fact, they remain satisfied throughout the course of the algorithm. With the *exception* of the vertex slackness conditions, the complementary slackness conditions are also satisfied at the outset. During the course of the algorithm, those complementary slackness conditions which are or which become satisfied remain thereafter satisfied. Moreover, each of the vertex conditions is eventually satisfied. When this happens the algorithm terminates and because all complementary slackness conditions are satisfied, the final matching is of maximum weight. Notice that if some vertex v does not satisfy its vertex slackness condition then $y_v > 0$ *and* v is a free vertex (throughout $\sum_v x(u, v)$ is zero or one).

The algorithm essentially consists of a reiterated step. Within each iteration there is an attempt to find an augmenting path (using the procedure *MAPS* described for the maximum-cardinality matching algorithm in section 5.2) in the subgraph $G' \subseteq G$ which consists of the edges $(u, v) \in E^*$ where:

$$E^* = \{(u, v) | y_u + y_v + \sum_{k:\,(u,v) \in Rk} z_k = w(u, v)\}$$

If an augmenting path is found, then it extends between two free vertices r and s for which:

$$y_r = y_s = W > 0$$

If we now augment along this strong augmenting path, interchanging edge rôles, then r and s are made to satisfy their vertex slackness conditions. Notice that we conveniently retain the term *strong augmenting path* (as we shall the term *neutral augmenting path*), even though it is not clear that the weight of the matching is increased. What *is* important is that this 'augmentation' causes two more vertices to satisfy their vertex slackness conditions. Because each edge (u, v) on the path belongs to the subgraph G', the edge slackness conditions also remain satisfied. Suppose that instead of finding a strong augmenting path in G' the search ends with a Hungarian tree. At this point changes are made to the dual variables. These changes can allow another edge or edges to be added to E^*, a pseudo-vertex to be expanded or cause the dual variable for some *outer* vertex to become zero. In the last case if the vertex in question is the root of the search tree then it now satisfies its vertex slackness condition; if it is not the root then the path from the root to the vertex is a neutral augmenting path and after augmentation along the path the root satisfies its vertex slackness condition. Notice that this augmentation frees the vertex v for which $y_v = 0$; however v still satisfies its vertex slackness condition simply *because* $y_v = 0$. If either of the first two cases occur (that is, edges are added to E^* or a pseudo-vertex is expanded) then the search for an augmenting path can continue from the same root. Eventually, this search will result in the root being made to satisfy its vertex slackness condition.

Fig. 5.9. The dual-variable changes procedure, *DVC*.

1. **for** all outermost vertices u labelled *outer* and all vertices u contained in an outermost blossom whose pseudo-vertex is labelled *outer* **do**

   $y_u \leftarrow y_u - \delta$
2. **for** all outermost vertices u labelled *inner* and all vertices u contained in an outermost blossom whose pseudo-vertex is labelled *inner* **do**

   $y_u \leftarrow y_u + \delta$
3. **for** all outermost blossoms R_k whose pseudo-vertices are labelled *outer* **do**

   $z_k \leftarrow z_k + 2\delta$
4. **for** all outermost blossoms R_k whose pseudo-vertices are labelled *inner* **do**

   $z_k \leftarrow z_k - 2\delta$

Changes to the dual variables involve a quantity δ as described in the dual variable changes procedure, *DVC*, of figure 5.9. Within lines 1 and 2 changes are made to y_u for vertices u either not contained in a blossom (outermost vertices) or to vertices u contained in an outermost blossom (but not contained in a more deeply nested blossom). We define δ to be a maximum such that the dual variables continue to provide a feasible solution to the dual problem. The dual constraints and non-negativity conditions continue to hold true if δ is determined by the δ-evaluation procedure, *DEV*, of figure 5.10. Notice that in statement 4 of that diagram, u and v are not contained in the same outermost blossom. If they were, then any changes to the dual variables would not affect the constraint:

$$y_u + y_v + \Sigma z_k \geqslant w(u, v)$$

because the changes of δ in each of y_u and y_v are offset by the change of 2δ in z_k for the outermost blossom containing both u and v.

Fig. 5.10. The δ-evaluation procedure, *DEV*.

---

1. $\delta_1 \leftarrow \frac{1}{2} \min_{R_k} \{z_k\}$

   where R_k is an outermost blossom whose pseudo-vertex is labelled *inner*.

2. $\delta_2 \leftarrow \min_{u} \{y_u\}$

   where u is any outermost vertex labelled *outer* or any vertex contained in an outermost blossom whose pseudo-vertex is labelled *outer*.

3. $\delta_3 \leftarrow \min_{(u, v)} \{y_u + y_v - w(u, v)\}$

   where u is any outermost vertex labelled *outer* or any vertex in an outermost blossom whose pseudo-vertex is labelled *outer* and v is an unlabelled vertex or is contained in an outermost blossom whose pseudo-vertex is unlabelled.

4. $\delta_4 \leftarrow \frac{1}{2} \min_{(u, v)} \{y_u + y_v - w(u, v)\}$

   where both u and v are each either outermost vertices labelled *outer* or vertices contained in different outermost blossoms whose pseudo-vertices are labelled *outer*.

5. $\delta \leftarrow \min \{\delta_1, \delta_2, \delta_3, \delta_4\}$

Consider the effect of the dual variable changes just described:

(*a*) If $\delta = \delta_1$ then some dual variable z_k becomes zero. We expand the associated pseudo-vertex back to its original odd-length circuit. The pseudo-vertex was labelled *inner* and so must have been the end-point of some edge in M. This edge therefore matches some vertex in the associated odd-length circuit. The other $2r_k$ vertices of the circuit can then be matched by adding circuit edges to M. Also when the pseudo-vertex is expanded, we can retain the existing labelling of vertices which defines T and (if the pseudo-vertex was of degree 2 in T) we can add to T the unique path around one side of the blossom which will keep T connected and alternating. In this case we label the vertices of the path *outer* and *inner* as appropriate.

(*b*) If $\delta = \delta_2$, then some dual variable y_u becomes zero. The path in the search tree from the root to u (if u is not the root) is an alternating path with the same number of edges in M as are not in M. If we interchange the edge rôles along this path then the root becomes matched (so satisfying the vertex slackness condition) and since $y_u = 0$, u also continues to satisfy the vertex slackness condition. If the y-variable for the root is zero then it has been made to satisfy the vertex slackness condition. In either case, we can now continue by growing a new search tree from another vertex v which is free and for which $y_v > 0$. If no such vertex exists then the algorithm stops.

(*c*) If $\delta = \delta_3$, then the associated edge can be added to E^* and the search for an augmenting path can be extended.

(*d*) If $\delta = \delta_4$, then the associated edge can be added to E^*. When the search for an augmenting path continues, this will result in the discovery of an odd cycle.

With the above explanation we can now present the Hungarian tree procedure *HUT* outlined in figure 5.11. When the procedure *MAPS* exits to *H*, *H* labels a call of *HUT*. Exits from *HUT* are either back to *MAPS*,

Fig. 5.11. The Hungarian tree procedure, *HUT*.

```
 1. DEV
 2. DVC
 3. if δ = δ1 then expand each outermost pseudo-vertex labelled
          inner which has a zero z-variable.
 4. if δ = δ2 and (the y-variable of the root T ≠ 0) then
          begin
 5.       Identify the alternating path P from the root to some
               vertex whose y-variable is zero.
 6.       Interchange edge rôles along P
          end
 7. if δ = δ3 or δ = δ4 then augment E*
 8. if δ = δ2 then begin
 9.                remove all inner and outer labels
10.                goto C
                   end
11. if δ = δ1 or δ = δ3 or δ = δ4 then goto M.
```

labelled *M*, or to a statement labelled *C*. This precedes *MAPS* and chooses the root of a new alternating tree. Notice that *HUT* subsumes the procedures *DEV* and *DVC* in lines 1 and 2. Line 3 expands pseudo-vertices as described in (*a*) above. The conditional statement of lines 4–6 deals with a (neutral) augmenting path as described in (*b*). Line 7 adds appropriate edges to E^* according to the prescriptions (*c*) and (*d*). If $\delta = \delta_2$ then the root of T is made to satisfy its vertex slackness condition so that a new root must be chosen for a new T. This is achieved through lines 8–10. In the other cases, $\delta = \delta_1$, δ_3 or δ_4, the current search tree can be continued with and so *MAPS* is recalled through line 11.

We are now in a position to present the maximum-weight matching algorithm which is outlined in figure 5.12. Line 1 initialises M to be the null matching. In line 2 the dual variables y_v are initialised, whilst the z-variable of each blossom is initialised in line 5 when the blossom is

Fig. 5.12. Maximum-weight matching algorithm.

```
    1. M ← ∅
    2. for all v ∈ V do
           y_v ← ½max {w(v_i, v_j)|(v_i, v_j) ∈ E}
C:  3. Choose a vertex v such that v is free and y_v > 0.
           If no such vertex exists goto L. Label v outer.
M:  4. MAPS(G′)
B:  5. Identify the blossom and shrink it. Label the resultant
           pseudo-vertex outer and assign zero to its z-variable.
           goto M.
H:  6. HUT
A:  7. Identify the (strong) augmenting path. Carry out
           augmentation by interchanging edge rôles. Remove
           all outer and inner labels.
           goto C.
L:  8. Expand all remaining pseudo-vertices in the final graph. Do
           this in the reverse order of their being found, inducing
           a maximum matching in each expanded blossom.
```

discovered. The general step of the algorithm is initiated in line 3 with the identification of a vertex whose vertex complementary slackness condition is not satisfied. Within line 4 the M-augmenting path search procedure grows an alternating tree rooted at this vertex and only using the edges E^* which define G'. As described earlier, exits from the procedure *MAPS* are to B if a blossom is found, to H if a Hungarian tree is discovered and to A on the discovery of a (strong) augmenting path. The statements of lines 5, 6 and 7 then result in a return to line 3 if the vertex complementary slackness condition of the root of the tree becomes satisfied or in a return to line 4 when the tree can be further developed. Eventually, all vertices satisfy their complementary slackness conditions and the algorithm terminates with line 8.

Our description of the maximum-weight algorithm very nearly amounts to its verification. As we indicated, growing an alternating tree from some root v will eventually result in its vertex complementary slackness condition becoming satisfied. As we have seen, this is always done so that M remains a feasible solution to the primal problem and so that the values y_v and z_k continue to provide feasible solutions to the dual problem. To complete verification of the algorithm we therefore only have to show that on termination, both the edge and odd-cardinality subset slackness conditions are satisfied. Notice that if an edge (u, v) is in M and not in a pseudo-vertex then $(u, v) \in E^*$. Also if (u, v) is contained within some pseudo-vertex then the value of $(y_u + y_v + \Sigma z_m)$ is unchanged by changes to the dual variables. Thus the edge slackness conditions are satisfied. Consider the z-variables. Any z_k can become positive only if it is contained within some pseudo-vertex. Whenever a pseudo-vertex is expanded a maximum matching is induced on the edges of the odd-circuit, so that on completion of the algorithm (when no pseudo-vertices remain) the odd-cardinality subset slackness conditions are satisfied.

It is easy to see that the maximum-weight matching algorithm is a polynomial time algorithm. We have, in our description, omitted details of the management of blossoms. Of course, a record of the nesting of blossoms has to be kept and continuously updated. Gabow[8] has detailed an $O(n^3)$ implementation of Edmonds' algorithm.

Fig. 5.13

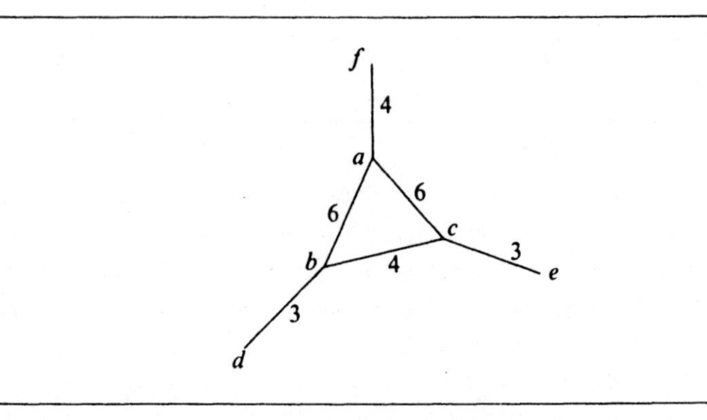

We now briefly describe an application of the algorithm to the graph of figure 5.13. Figure 5.14(*a*) shows additions to E^*, additions and deletions to M and variations in the variable z_B as the algorithm proceeds. In figure 5.14(*b*) we plot changes to the y-variables. In fact the vertices *a*, *c*, *d*, *e* and *f*, respectively, are taken in turn as the root of a new search tree at each iteration of the algorithm starting at line 3 of figure 5.12.

Fig. 5.14

(*a*)

	(*a*)	(*c*)			(*d*)	(*e*)
Variations to M	Add (a, b)	Remove (a, b)			Add (d, b)	Remove (a, c) Add (f, a) and (c, e)
Additions to E^*		(b, c)		(d, b) (c, e)		
Z_B			0	4		0

(*b*)

	y_a	y_b	y_c	y_d	y_e	y_f
	3	3	3	3	3	3
(*a*)						
(*c*)	4	2	2	3	3	3
	2	0	0	3	3	3
(*d*)						
(*e*)	4	2	2	1	1	3
	3	1	1	0	0	3

The following cryptic notes indicate what happens within each iteration:

choice of a (a is labelled *outer*)

MAPS(G'): (a, b) is explored. b is free and unlabelled so that:

$T \leftarrow \underset{(outer)}{a} \text{———} b$

and there is an exit to A

A: (a, b) is the (strong) augmenting path so that (a, b) is added to M, the label *outer* is removed from a.
Exit to C.

choice of c (c is labelled *outer*)

MAPS(G'): (c, a) is explored:

$T \leftarrow \underset{outer}{c} \text{———} \overset{inner}{a} \sim\sim \underset{outer}{b}$

(edges in M are denoted by wavy line).
No edge is explorable from b or c.
Exit to H.

HUT:*DEV*: $\delta_1 = \infty$, $\delta_2 = 3$ (vertices b and c)
$\delta_3 = 3$ (edges (b, d) and (c, e)), $\delta_4 = 1$ (edge (b, c))
Hence $\delta = \delta_4 = 1$.

DVC: $y_a \leftarrow 3-1 = 2$, $y_b \leftarrow 3-1 = 2$
(b, c) joins E^*. Exit to M.

MAPS(G'): (b, c) is explored; since c is labelled *outer* a blossom has been found consisting of the edges (c, a), (a, b) and (b, c). Exit to B.

B: The blossom is shrunk and the pseudo-vertex, which we denote by B is labelled *outer*. (a, b) is removed from M. $z_B \leftarrow 0$. Exit to M.

MAPS(G'): No edge is explorable from B. Exit to H.

HUT:*DEV*: $\delta_1 = \infty$, $\delta_2 = 2$ (vertices b and c)
$\delta_3 = 2$ (edges (d, b) and (c, e)), $\delta = \infty$
Hence $\delta = \delta_2 = \delta_3 = 2$.

DVC: $y_a \leftarrow 4-2 = 2$, $y_b \leftarrow 2-2 = 0$
$y_c \leftarrow 2-2 = 0$ (the y-variable of the root is zero)
$z_B \leftarrow 0+2.2 = 4$

(d, b) and (c, e) join E^*
All *outer* and *inner* labels are removed.
Exit to C.

choice of d (d is labelled *outer*)

MAPS(G'): (d, b) is explored. b is free and unlabelled so that:

$T \leftarrow d$ —— B
outer

Exit to A.

A: (d, b) is the augmenting path so that (d, b) is added to M. The label *outer* is removed from d. Exit to C.

choice of e (e is labelled *outer*)

MAPS(G'): (e, c) is explored so that:

inner
$T \leftarrow e$ —— . ∿∿∿ d
outer B *outer*

No edges are explorable from d or e.
Exit to H.

HUT:*DEV*: $\delta_1 = 2$ (vertex B), $\delta_2 = 3$ (vertices d and e)
$\delta_3 = \infty$, $\delta_4 = \infty$
Hence $= \delta_1 = 2$

DVC: $y_a \leftarrow 2+2 = 4$, $y_b \leftarrow 0+2 = 2$, $y_c \leftarrow 0+2 = 2$
$y_d \leftarrow 3-2 = 1$, $y_e \leftarrow 3-2 = 1$, $z_B \leftarrow 4-2.2 = 0$

The blossom associated with the pseudo-vertex B is expanded. The even-length side of the blossom is interpolated into T with appropriate labelling of vertices. (a, c) is added to M:

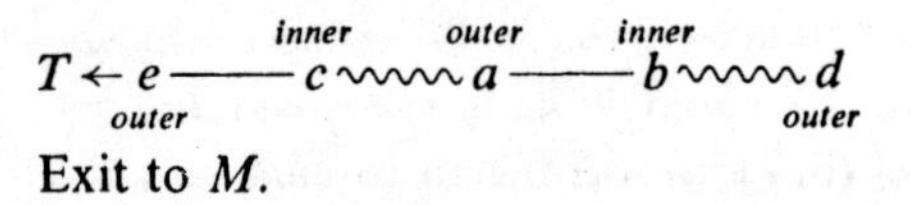

Exit to M.

MAPS(G'): No explorable edges exist from e, a or d.
Exit to H.

HUT:*DEV*: $\delta_1 = \infty$, $\delta_2 = 1$ (vertices d and e)
$\delta_3 = 3$ (edge (a, f)), $\delta_4 = \infty$
Hence $\delta = \delta_2 = 1$.

DVC: $y_a \leftarrow 4-1 = 3$, $y_b \leftarrow 2-1 = 1$, $y_c \leftarrow 2-1 = 1$
$y_d \leftarrow 1-1 = 0$, $y_e \leftarrow 1-1 = 0$ (The y-variable of the root of T becomes zero.)

All *inner* and *outer* labels are removed.
Exit to C.

choice of f (f is labelled *outer*)

MAPS(G'): (f, a) is explored:

$$T \leftarrow \underset{outer}{f} \text{———} \overset{inner}{a} \rightsquigarrow \underset{outer}{c}$$

(c, e) is explored:

$$T \leftarrow \underset{outer}{f} \text{———} \overset{inner}{a} \rightsquigarrow \overset{outer}{c} \text{———} e$$

e is free and unlabelled. Exit to A.

A: The augmenting path is $((f, a), (a, c), (c, e))$ hence (a, c) is removed from M while (f, a) and (c, e) are added to M. All *inner* and *outer* labels are removed. Exit to C.

No free vertices exist, jump to L. No unexpanded pseudo-vertices and so the algorithm stops with:

$$M = ((a, f), (b, d), (c, e))$$

5.4. Summary and references

The maximum-cardinality matching and maximum-weight matching algorithms for general graphs, which form the central content of this chapter, are essentially due to the pioneering work of Edmonds. Whilst Edmonds' work provides the guiding principles for these algorithms, we have exercised considerable licence in the rather particular presentations of this chapter. As we have seen, efficient algorithms exist for the maximum-cardinality and for the maximum-weight matching problems and also for the question of determining whether or not a graph contains a perfect matching.

Edge coverings, which we define in exercise 5.10, provide problems which are similar to those provided by matchings. In exercise 5.10, we indicate how the problem of finding a *minimum-cardinality covering* can be solved in polynomial time using the maximum-cardinality matching algorithm. There is not, unfortunately, a similar relationship between the maximum-*weight* matching problem and the problem of finding a minimum-*weight* covering. White[13] has, however, described a polynomial time minimum-weight covering algorithm which is in the same spirit as the maximum-weight matching algorithm of section 5.3.

Chapter 5 of Minieka[14] and chapters 5 and 6 of Lawler[15] are recommended reading for the material presented in this chapter.

The exercises that follow provide some illustrations of how matching problems can arise naturally in a variety of guises.

[1] Edmonds, J. 'An introduction to matching', Lecture notes. *Engineering Summer Conference*, the University of Michigan, Ann Arbor, USA (1967).

[2] Edmonds, J. 'Paths, trees and flowers', *Can. J. Math.*, **17**, 449–67 (1965).

[3] Micali, S. & Vazirani, V. V. 'An $O(\sqrt{n} \cdot |E|)$ algorithm for finding maximum matchings in general graphs', *21st Annual Symposium on Foundations of Computer Science*, Syracuse, NY, pp. 17–27 (NY, USA, IEEE 1980).

[4] Tutte, W. T. 'The factorisation of linear graphs', *J. London Math. Soc.*, **22**, 107–11 (1947).

[5] Lovász, L. 'Three short proofs in graph theory', *J. Combinatorial Theory, B*, **19**, 111–13 (1975).

[6] Kuhn, H. W. 'Variants of the Hungarian method for assignment problems', *Naval Res. Logist. Quart.*, **3**, 253–8 (1956).

[7] Edmonds, J. & Johnson, E. 'Matching: A well-solved class of integer linear programs', *Combinatorial Structures and Their Applications*, Gordon & Breach, NY, pp. 89–92 (1970).

[8] Gabow, H. 'An efficient implementation of Edmonds' maximum matching algorithm', *Technical Report 31*, Stanford University Computer Science Dept (June 1972).

[9] Hall, P. 'On representatives of subsets', *J. London Math. Soc.*, **10**, 26–30 (1935).

[10] Fujii, M. *et al.* 'Optimal sequencing of two Equivalent Processors', *SIAM J. Appl. Math.*, **17**, 784–9 (1969).

[11] Coffman, E. G. & Graham, R. L. 'Optimal scheduling for two-processor systems', *Acta Informatica*, **1**, 200–13 (1972).

[12] Gale, D. & Shapley, L. S. 'College admissions and the stability of marriage', *Amer. Math. Monthly*, **69**, 9–14 (1962).

[13] White, L. J. A parametric study of matchings and coverings in weighted graphs, PhD Thesis, University of Michigan (1967).

[14] Minieka, E. *Optimisation Algorithms for Networks and Graphs*, Marcel Dekker, NY (1978).

[15] Lawler, E. *Combinatorial Optimisation*: *Networks and Matroids*, Holt, Rinehart and Winston (1976).

EXERCISES

5.1. A multinational army has a tank corps. Each tank requires a crew of two who speak a common language. Each possible crew member generally speaks more than one language. How might the problem of maximising the number of crews be reduced to the problem of finding a maximum-cardinality matching for a graph in which each vertex represents a possible crew member?

5.2. A theatrical agent receives offers of employment for some of his actors from a number of theatrical impresarios. Each impresario wishes to employ just one actor and, in an attempt to best meet his requirements, he offers different rates of pay for the actors he is offered.

How might the problem of maximising the agent's income (if he receives a fixed percentage of his actors' incomes) be reduced to the problem of finding a maximum-weight matching for a graph in which each vertex represents either an actor or an impresario?

5.3. The following is a statement of Hall's[9] theorem. If $G = (V, E)$ is a bipartite graph with bipartition (V', V''), then G has a complete matching of V' onto V'' if and only if:

$$|N(V_i')| \geqslant |V_i'|, \quad \text{for all} \quad V_i' \subseteq V'$$

where $N(V_i')$ is the set of vertices adjacent to V_i'.

Obtain Hall's theorem from theorem 5.2.

(Suppose that $|V|$ is even. We can construct G^* by adding to G an edge joining every pair of vertices in V''. Show that G has a matching, with every vertex of V' an end-point of an element of the matching, if and only if G^* has a perfect matching. Hall's theorem then follows naturally. For $|V|$ odd, a simple modification to the proof is required.)

5.4. *Job assignments*

An employer wishes to fill i vacancies with pretrained skilled labour. An employment agency provides a list of j potential employees, each having been trained for one or more of the vacancies. Using Hall's theorem (exercise 5.3) how might the prospects of:

(*a*) filling all the vacancies,

and

(*b*) employing all the candidates,

be judged?

Does Hall's theorem provide an efficient way to answer these questions? How might a maximum number of vacancies be filled in polynomial time?

5.5. *The marriage problem*

In a certain community every boy knows exactly k girls and every girl knows exactly k boys. Show that every boy can marry a girl he knows and vice versa.

(Construct a k-regular bipartite graph in which each edge signifies that a boy (represented by one end-point) knows a girl (represented by the other end-point). The problem then reduces to showing that any k-regular bipartite graph, $k > 0$, has a perfect matching. Show that the number of boys must equal the number of girls and that the graph satisfies Hall's theorem (exercise 5.3). The result then follows.)

5.6. $G = (V, E)$ is any 3-regular graph without cut-edges. Let

$$G_i = (V_i, E_i), \; i = 1, 2, \ldots, k$$

be the odd components (see theorem 5.2) of $(G-V')$, $V' \subset V$. If m_i is the number of edges with one end-point in G_i and the other in V', show that m_i is odd and that $m_i \geqslant 3$. Justify the following:

$$\Phi(G-V') = k \leqslant \frac{1}{3}\sum_{i=1}^{k} m_i \leqslant \tfrac{1}{3}\sum_{v\in V'} d(v) = |V'|$$

thus proving, by theorem 5.2, that G has a perfect matching.

5.7. *Shortest time scheduling for two processors*

A complicated task can be broken down into a number of subtasks, s_i, $i = 1, 2, \ldots, n$, each requiring a unit of processing time. Two processors are available and can operate simultaneously. There exists a partial ordering '$<$' for the s_i, such that $s_i < s_j$ means that s_i must be completed before s_j. G is a digraph in which each vertex represents some s_i and there is an edge (s_i, s_j) for each relation $s_i < s_j$. An undirected graph $G^* = (V, E^*)$ is constructed as follows. G^* has the same vertex set as G and $(s_i, s_j) \in E^*$ if and only if there is no directed path from s_i to s_j or from s_j to s_i in G. Such a construction is shown below. Justify the statement that if M is a maximum-cardinality matching in G^*, then a lower bound in the computation time for the overall task is given by $(n-|M|)$.

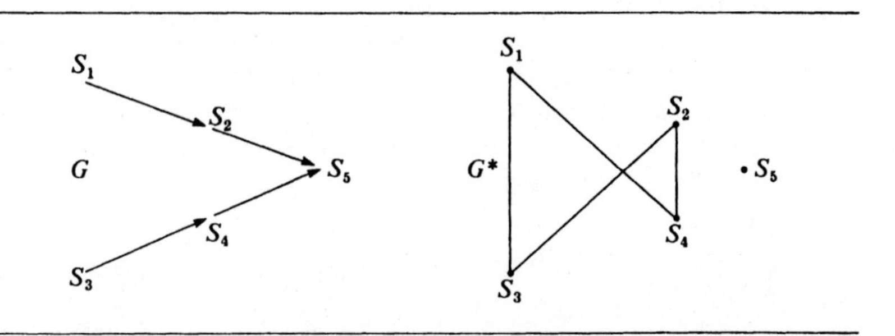

(Such a matching for the problem is said to be *feasible* if it describes a possible scheduling sequence. For example, matching $\{(s_1, s_3), (s_2, s_4)\}$ describes a feasible schedule: (s_1 and s_3) being executed simultaneously, followed by (s_2 and s_4), and finally s_5 is executed. However, the matching $\{(s_1, s_4), (s_2, s_3)\}$ is not feasible. Fujii *et al.*[10] have shown that a feasible matching always exists which is of maximum cardinality and that this can be found in $O(n^3)$-time. See also Coffman & Graham.[11])

5.8. *The stable marriages problem*

In a community of n men and n women each person ranks those of the opposite sex according to his or her preference for a marriage partner. The problem is to marry off all members of the community in such a way that the set of marriages is stable. The set is unstable if a man and a woman exist who are not married to each other but prefer each other to their actual mates.

Gale & Shapley[12] have described the following algorithm to solve the problem:

To start, let each boy propose to his favourite girl. Each girl who receives more than one proposal rejects all but her favourite from amongst those who have proposed to her. However, she does not accept him yet, but keeps him on a string to allow for the possibility that someone better may come along later.

We are now ready for the second stage. Those boys who were rejected now propose to their second choices. Each girl receiving proposals chooses her favourite from the group consisting of the new proposers and the boy on her string, if any. She rejects all the rest and again keeps the favourite in suspense.

We proceed in the same manner. Those who are rejected at the second stage propose to their next choices, and the girls again reject all but the best proposal they have had so far.

Eventually...every girl will have received a proposal, for as long as any girl has not been proposed to there will be rejections and new proposals, but since no boy can propose to the same girl more than once, every girl is sure to get a proposal in due time. As soon as the last girl gets her proposal the 'courtship' is declared over, and each girl is now required to accept the boy on her string.

Provide brief justification for the following claims:

(*a*) The algorithm provides a stable set of marriages.

(*b*) The algorithm also works if the number of males does not equal the number of females.

(*c*) The algorithm has complexity $O(n^2)$.

(*d*) The algorithm rejects men only from women that they could not possibly be married to under any stable matching. That is, that any man is at least as well-off as he would be under any other stable marriage. The algorithm is called *man-optimal* for this reason. A *woman-optimal* set of marriages is, of course, obtained by getting the women to propose to the men.

5.9. By simplifying the algorithm described in section 5.3, produce a maximum-weight matching algorithm specifically for bipartite graphs.

5.10. A *covering* C is any set of edges such that any vertex of the graph is an end-point of (at least) one edge of C. A *minimum-cardinality covering* is a covering with the smallest possible number of edges.

(*a*) A salesman in educational toys has a selection of geometrical shapes (cubes, pyramids and so on), each of which is manufactured in a range of colours. He wishes to carry with him a minimum number of objects so that each colour and each shape is represented at least once. Justify the following statement. The number of objects he must carry is equal to the number of elements in a minimum-cardinality covering in the graph where each shape and each colour are individually represented by a single vertex and there is an edge joining a shape vertex to a colour vertex if that shape is manufactured in that colour.

(*b*) Let M be a maximum cardinality matching and C be a minimum cardinality covering of $G = (V, E)$. Now construct:

(i) a covering C' from M by adding to M, for every unmatched vertex v, one edge incident with v,

and

(ii) a matching M' from C by removing from C, for every *overcovered* vertex v (that is, v is the end-point of more than one edge of C) all but one edge incident with C.

Show that:

$$|C'| = |V| - |M|$$

and that

$$|M'| = |V| - |C|$$

Hence deduce that C' is a minimum-cardinality covering and that M' is a maximum-cardinality matching. Thus the problem of finding a minimum-cardinality covering can be solved essentially by the maximum-cardinality matching algorithm of section 5.2.

6

Eulerian and Hamiltonian tours

In this chapter we concentrate on two fundamental ways of traversing a graph. In historical terms these represent perhaps the oldest areas of inquiry in graph theory. The first concerns paths or circuits in which every edge is used precisely once. These are called *Eulerian* after the Swiss mathematician L. Euler. He published in 1736 (see exercise 6.2) what is often referred to as the first paper in graph theory. The second way of traversing a graph of interest to us involves visiting each vertex precisely once. These paths or circuits are called *Hamiltonian* after the English mathematician W. R. Hamilton who studied them (*circa* 1856) in connection with a game of his invention (see exercise 6.3).

In connection with Eulerian and Hamiltonian paths and circuits, the word *tour* will mean either a path or a circuit. We shall be interested in characterising graphs that contain either Eulerian or Hamiltonian tours. Also, we shall be investigating the well-known and related problems of the Chinese postman and of the travelling salesman.

6.1 Eulerian paths and circuits

A postman delivers mail every day in a network of streets. In order to minimise his journey he wishes to know whether or not he can traverse this network and return to his depot without walking the length of any street more than once. This problem concerns the existence or otherwise of an Eulerian circuit of the corresponding graph. If one exists then he may wish to know how many others do in order to vary the otherwise tedious routine. We shall see in this section just how questions of this type may be answered.

Fig. 6.1. (*a*) An Eulerian circuit of G_1. (*b*) An Eulerian path of G_2.

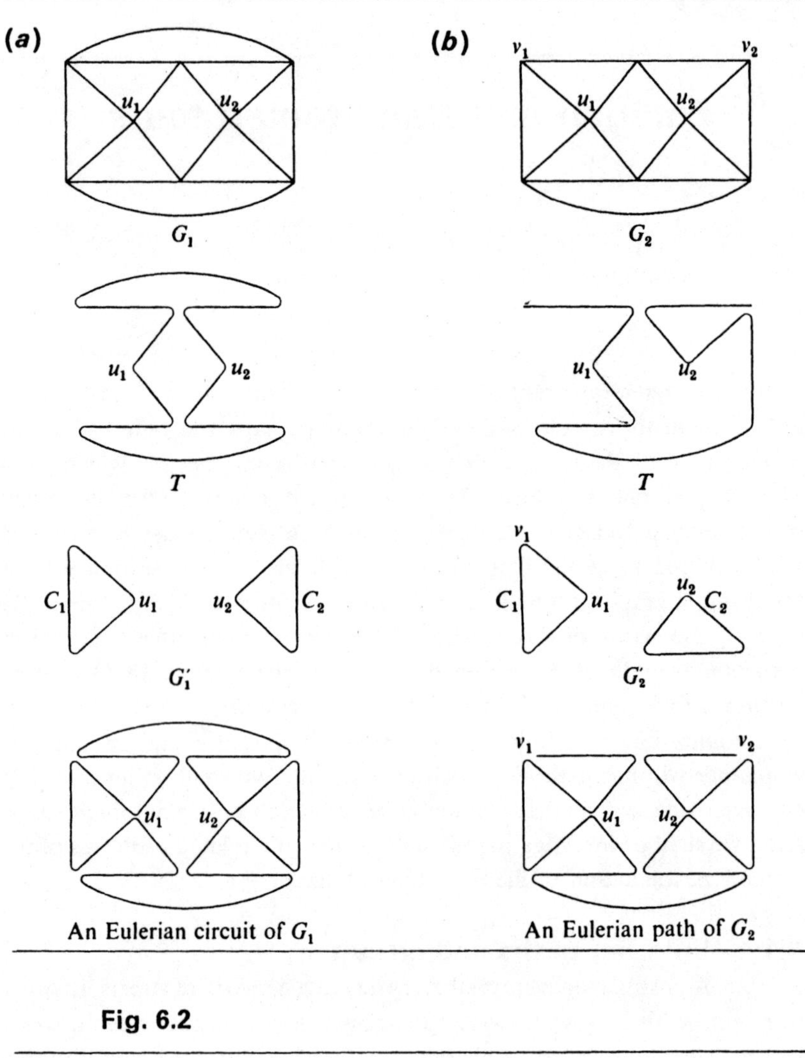

Fig. 6.2

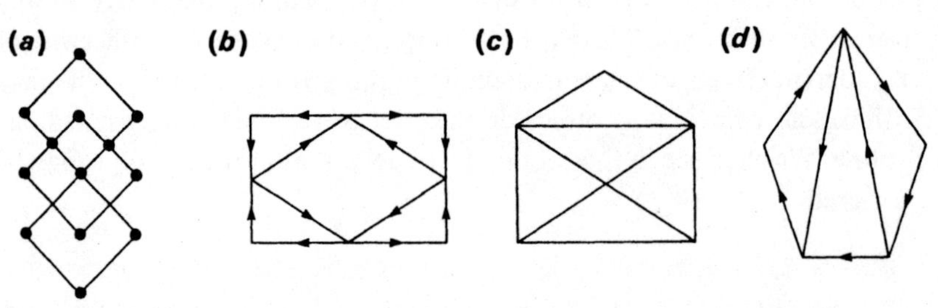

6.1.1. Eulerian graphs

An *Eulerian graph* is an undirected graph, or a digraph, which contains an Eulerian circuit. Of course, for a digraph each edge of the circuit can only be traversed as it is directed. The following theorem determines whether or not an undirected graph is Eulerian or contains an Eulerian path.

Theorem 6.1. An undirected multi-graph G, has an Eulerian circuit (or path) if and only if it is connected and the number of vertices with odd-degree is 0 (or 2).

Proof. The conditions are clearly necessary because if an Eulerian tour exists then G must be connected and only the vertices at the ends of an Eulerian path can be of odd-degree.

To show sufficiency we use induction on the number of edges $|E|$. The theorem is trivially true for $|E| = 2$. Let G have $|E| > 2$ edges, and let it satisfy the conditions of the theorem. If G contains two vertices of odd-degree, we denote them by v_1 and v_2. Consider tracing a tour T from a vertex v_i ($= v_1$ if there are vertices of odd-degree). We trace T leaving each new vertex encountered by an unused edge until a vertex v_j is encountered for which every incident edge has been used. If G contains no vertices of odd-degree then it must be the case that $v_i = v_j$, otherwise it must be the case that $v_j = v_2$. Suppose that T does not use every edge of G. If we remove from G all those edges that have been used, then we are left with a, not necessarily connected, subgraph G'. G' only contains vertices of even-degree. By the induction hypothesis each component of G' contains an Eulerian circuit. Since G is connected, T must pass through at least one vertex in each component of G'. An Eulerian tour can then be constructed for G by inserting into T, at one such vertex for each component of G', an Eulerian circuit for that component. ■

Figure 6.1 illustrates the construction of an Eulerian circuit for G_1 in (*a*) and the construction of an Eulerian path for G_2 in (*b*) both according to the prescription of the above proof. In both cases the original graph, less the edges of T, forms a graph with two components, C_1 and C_2. The vertices u_1 and u_2 indicate the points where the Eulerian circuits of these components are inserted into T.

The following corollary applies to digraphs. Its proof exactly parallels that for theorem 6.1.

Corollary 6.1. A digraph is Eulerian if and only if it is connected and is balanced. A digraph has an Eulerian path if and only if it is connected and the degrees of its vertices satisfy:

$$d^+(v) = d^-(v) \quad \text{for all} \quad v \neq v_1 \text{ or } v_2$$
$$d^+(v_1) = d^-(v_1)+1$$
$$d^-(v_2) = d^+(v_2)+1$$

Given theorem 6.1 and its corollary check that in figure 6.2, graphs (*a*) and (*b*) have Eulerian circuits while (*c*) and (*d*) have Eulerian paths but not circuits.

6.1.2. Finding Eulerian circuits

Theorem 6.1 and its corollary describe algorithms to find Eulerian circuits in graphs and digraphs. We describe here algorithms which construct Eulerian circuits directly in the sense that they do not proceed by the repeated addition of subcircuits.

We first describe an algorithm which is applicable to undirected graphs. It will become evident that the same algorithm may be utilised for digraphs. However, in the case of digraphs, it will be useful for other purposes to describe a second algorithm. The first algorithm then is outlined in figure 6.3. Given an undirected Eulerian graph $G = (V, E)$, the algorithm traces an Eulerian circuit during which CV denotes the current vertex being visited, E' denotes the set of edges already traced and EC is a list of vertices ordered according to the sequence in which they have been visited. Also $A(v)$ denotes the adjacency list of the vertex v within the graph $(G-E')$. The first vertex visited is w. When the circuit has been traced as far as CV, the conditional statement starting at line 5 chooses the next vertex v that shall be visited. This is done so that if (CV, v) is not the only edge incident with CV in $(G-E')$, then (CV, v) is not a cut-edge of $(G-E')$. Such a choice is always possible because, as we shall see in the next theorem, there can only ever be at most one cut-edge of $(G-E')$ incident with CV. This important fact also means that the search for v in line 6 will be restricted to checking whether or not (CV, v'), where v' is the first vertex in $A(CV)$, is a cut-edge of $(G-E')$. If it is not, then v' becomes v, otherwise the second vertex in $A(CV)$ becomes v. Before validifying the algorithm we establish its complexity according to the implementation of figure 6.3.

The body of the while statement, lines 5–11, is executed $|E|$ times, once for each successive edge traced in the Eulerian circuit. Within each execution we need, in line 6, to determine whether or not a particular edge of $(G-E')$ is a cut-edge. It is easy to do this in $O(|E-E'|)$-time by searching for a path in $G-(E' \cup (v_i, v_j))$. We simply tag neighbours of V_i, then repeatedly tag untagged neighbours of tagged vertices until the process cannot proceed further. If, finally, v_j remains untagged then (v_i, v_j) must be a cut-edge of $(G-E')$. Thus overall we have an $O(|E|^2)$ implementation. We can in fact with little trouble find Eulerian circuits in $O(|E|)$-time. See, for example, exercise 6.5.

Fig. 6.3

1. $EC \leftarrow [w]$
2. $CV \leftarrow w$
3. $E' \leftarrow \varnothing$
4. **while** $|A(w)| > 0$ **do**
    begin
5. **if** $|A(CV)| > 1$ **then**
6. find a vertex $v \in A(CV)$ such that (CV, v) is not a cut-edge of $(G-E')$
7. **else** let the vertex in $A(CV)$ be denoted by v
8. delete v in $A(CV)$ and CV in $A(v)$
9. $E' \leftarrow E' \cup \{(CV, v)\}$
10. $CV \leftarrow v$
11. add CV to the tail of EC
    end
12. Output EC

Theorem 6.2. The algorithm of figure 6.3 finds an Eulerian circuit EC of an undirected graph $G = (V, E)$.

Proof. We first show that the choice of the next vertex v, within the conditional statement starting at line 5, is always possible. Having arrived at CV ($\neq w$) it must be that $|A(CV)| > 0$ and that $|A(CV)|$ is odd because $d(CV)$ is even. If $|A(CV)| = 1$ then the next vertex is uniquely determined by the else clause at line 7. However, if $|A(CV)| > 1$ then at most one edge incident to CV can be a cut-edge. We can see this by noting that any component of the graph $(G-E')$ attached to CV by a cut-edge must contain a vertex of odd-degree in $(G-E')$. Suppose this were not so. Then every vertex of the component will be of even-degree so that the sum of these degrees will be even. However, the sum of these degrees is odd because each edge of the component adds two to the sum except for the single cut-edge attaching it to CV. Now there are precisely two vertices of odd-degree in $(G-E')$, namely, w and CV. Hence, there can only be at most one cut-edge of $(G-E')$ adjacent to CV ($\neq w$). Suppose now that $CV = w$. If $A(w) > 0$, as required by line 4, then no edge incident to CV is a cut-edge of $(G-E')$. This follows by noting that when the Eulerian circuit revisits w, then every vertex of $(G-E')$ is of even-degree. But by a previous argument if a cut-edge attaches a component of $(G-E')$ to w, then this component would contain at least one vertex of odd-degree.

Thus a next vertex can always be chosen as the algorithm requires within line 6. According to line 4 the process stops when $|A(w)| = 0$. At this stage an Eulerian circuit will have been traced. Otherwise, before reaching w, some other vertex u with $|A(u)| > 1$ must have been left along a cut-edge of $(G-E')$. ■

We now describe an algorithm specifically appropriate for digraphs. It constructs an Eulerian circuit starting with a spanning out-tree of the digraph. This construction will be of interest to us again when we come to count the Eulerian circuits of a graph in section 6.2.1.

Before describing the algorithm we show that the reverse construction is possible. That is, given an Eulerian circuit of a digraph we can construct a spanning out-tree. Starting at an arbitrary vertex u, we trace the Eulerian circuit and, for each vertex except u, we identify the first edge incident to the vertex. According to theorem 6.3 this set of $(n-1)$ edges constitutes a spanning out-tree of the digraph.

Theorem 6.3. The subgraph of an Eulerian digraph G constructed according to the above rule is a spanning out-tree of G rooted at u.

Proof. We denote the subgraph by T. By the construction rule we see that within T $d^-(u) = 0$, while for every vertex $v \neq u$ $d^-(v) = 1$. Then since T has $(n-1)$ edges we need just show that T is acyclic.

Suppose that T contains a cycle. As edges are added to T according to the construction rule, let (v_i, v_j) be the first edge that completes a circuit in T. Clearly, $v_j \neq u$. Since (v_i, v_j) completes a circuit, v_j has been visited previously in tracing the Eulerian path. Thus (v_i, v_j) cannot be an initial entry to v_j and so would not be included in T. This is a contradiction and so T is acyclic. ■

Figure 6.4 shows a digraph G and an Eulerian circuit C. It also shows a spanning out-tree T constructed according to the rule described for the previous theorem.

Fig. 6.4

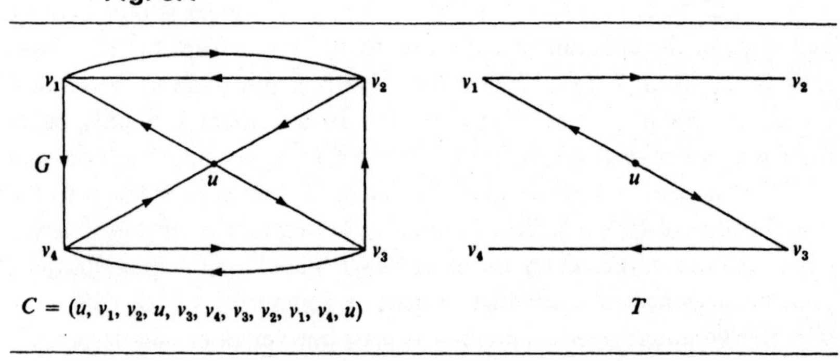

By a proof similar to that for theorem 6.3 it is easy to see that a spanning in-tree T, rooted at u, for an Eulerian digraph G can be constructed as follows. Starting at u trace an Eulerian circuit adding to T those edges which correspond to final exits from each of the $(n-1)$ vertices other than u. It is also clear that we can construct a spanning-tree for an undirected graph by these methods by temporarily regarding the edges to be directed in the same sense as an Eulerian circuit is traced.

We now return to the algorithm which finds an Eulerian circuit of a digraph given a spanning out-tree T. This is embodied in theorem 6.4 which is clearly the converse of theorem 6.3.

Theorem 6.4. If G is a connected, balanced digraph with a spanning out-tree T rooted at u, then an Eulerian circuit can be traced in *reverse* direction as follows:

(*a*) The initial edge is any edge incident to u.
(*b*) Subsequent edges are chosen so as to be incident to the current vertex and such that:
  (i) no edge is traversed more than once,
  (ii) no edge of T is chosen if another edge is still available.
(*c*) The process stops when a vertex is reached which has no unused edges incident to it.

Proof. Since G is balanced, the path traced by the above rules can only terminate at u. However, suppose that this circuit does not contain an edge (v_i, v_j) of G. Now, G is balanced and so v_i must be the final vertex of some other unused edge (v_k, v_i). We can take (v_k, v_i) to be an edge of T since such an edge incident to v_i will not have been used because of rule (b(ii)). We can now by a sequence of similar edges follow a directed path backwards to u. Because G is balanced we should then find an edge incident to u which has not been used in the circuit. But this contradicts rule (c). Hence the circuit must be Eulerian. ∎

The algorithm of theorem 6.4 can be executed in $O(|E|)$-time as can be readily seen by an inspection of figure 6.5. Within that diagram the first statement indicates the initial task of finding a spanning out-tree. This tree is represented by the set of boolean variables $\{T(e) | e \in E\}$. It is easy to see that if an Eulerian digraph is subjected to the *DFS* algorithm of figure 1.15, then it will find a spanning out-tree. Since that algorithm operates in $O(\max(n, |E|))$-time and since for an Eulerian digraph $|E| \geqslant n$, the first task of figure 6.5 required $O(|E|)$-time. For each vertex v, the for statement at line 5 constructs A_v, which is a list of the vertices v_i such that $(v_i, v) \in E$. Moreover, that edge incident to v ($\neq u$) and which is an edge

Fig. 6.5. An O($|E|$)-algorithm to find an Eulerian circuit of a digraph $G = (V, E)$.

```
1.  Find a spanning out-tree T of G = (V, E) rooted at u,
        representing it by the assignments:
        T((v_i, v_j)) ← if (v_i, v_j) ∈ T then true else false
2.  for every vertex v ∈ V do
            begin
3.          A_v ← ∅
4.          I(v) ← 0
            end
5.  for every edge (v_i, v_j) ∈ E do
        if T((v_i, v_j)) then add v_i to the tail of A_{v_j}
            else add v_i to the head of A_{v_j}
6.  EC ← ∅
7.  CV ← u
8.  while I(CV) ≤ d⁻(CV) do
        begin
9.      add CV to the head of EC
10.     I(CV) ← I(CV)+1
11.     CV ← A_CV(I(CV))
        end
12. Output EC
```

of the spanning out-tree is arranged to be the last edge in the list A_v. Clearly, all the A_v are constructed within O($|E|$)-time. The for statement beginning at line 2 provides an initial assignment of the empty list to each A_v and zero to each $I(v)$ and does this within O(n)-steps. $I(v)$ is used as an index to the list A_v and CV denotes the current vertex being visited. Thus in line 11 $A_{CV}(I(CV))$ means the $I(CV)$th element in the list A_{CV}. EC eventually lists the vertices of the Eulerian circuit in the order in which they are visited. The circuit is traced within the while statement starting at line 8. Rule (*b*) of theorem 6.3 is ensured by the construction of the A_v and the incrementation of the $I(v)$ in line 10. Rule (*c*) is taken care of by the condition of line 8 and rule (*a*) is ensured by the assignment of line 7. Since the while body, lines 9–11, is executed once for each edge in the Eulerian circuit, the while statement requires O($|E|$)-time. Thus overall we have an O($|E|$)-algorithm.

Figure 6.6 illustrates an application of the algorithm of figure 6.5. The spanning out-tree produced by the first stage of the algorithm consists of those edges e for which it is shown that $T(e)$ = **true**. For each $v \in V$, the lists A_v are also shown. The table then shows for each iteration of the while loop, lines 9–11 of figure 6.5, the values of CV and each $I(v)$. The final state of EC which is shown indicates which Eulerian circuit is found.

Fig. 6.6. An application of the algorithm in figure 6.5.

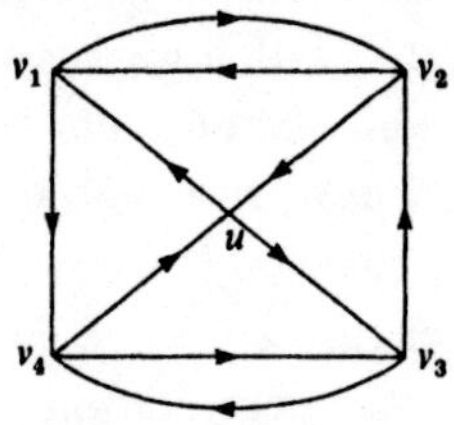

$A_{v_1} = [v_2, u]$ $T((v_1, v_2)) =$ **true**

$A_{v_2} = [v_3, v_1]$ $T((u, v_1)) =$ **true**

$A_{v_3} = [v_4, u]$ $T((u, v_3)) =$ **true**

$A_{v_4} = [v_1, v_4]$ $T((v_3, v_4)) =$ **true**

$A_u = [v_2, v_4]$ for all v, $d^-(v) = 2$

Iteration	CV	$I(u)$	$I(v_1)$	$I(v_2)$	$I(v_3)$	$I(v_4)$
0	u	0	0	0	0	0
1	v_2	1	0	0	0	0
2	v_3	1	0	1	0	0
3	v_4	1	0	1	1	0
4	v_1	1	0	1	1	1
5	v_2	1	1	1	1	1
6	v_1	1	1	2	1	1
7	u	1	2	2	1	1
8	v_4	2	2	2	1	1
9	v_3	2	2	2	1	2
10	u	2	2	2	2	2
11	—	3	2	2	2	2

$EC = [u, v_3, v_4, u, v_1, v_2, v_1, v_4, v_3, v_2, u]$

6.2 Postman problems

We now consider problems of the type posed by the postman in the opening paragraph of section 6.1. The question of whether or not the postman can traverse his network of streets, starting and finishing at the depot and traversing each street exactly once, can now be easily answered. If the streets can be traversed in either direction then the Eulerian test of theorem 6.1 provides the answer, whilst if the streets are one-way traversable then we can refer to corollary 6.1. If the network of streets is *not* Eulerian then we can naturally ask a further question. How can we find a shortest circuit for the postman which visits each street at least once? Here we associate a length with each street so that the associated graph is weighted. Because of the origin of an early paper[1] describing it, this problem is called the *Chinese postman problem*. We shall devote most of this section to solving this problem both for undirected and for directed graphs. Before coming to that, however, we deal with the problem of counting the number of distinct Eulerian circuits in an Eulerian graph.

6.2.1. Counting Eulerian circuits

For digraphs, on the basis of theorem 6.4, we can count the number of distinct Eulerian circuits associated with a given spanning out-tree by considering the choice of edges available at each vertex as the circuit is traced. Let the out-tree be rooted at u. In counting circuits we must fix the final edge that is to be traced backwards from u in order to avoid multiple counting. Otherwise each circuit would be counted $d^-(u)$ times, any two counts differing only by a cyclic permutation of the edges. Also, the choice of edge to be traced backwards from any other vertex is restricted in that the edge associated with the spanning out-tree must be traced last. An Eulerian circuit encounters any vertex v, $d^-(v)$ times. On the first occasion the circuit has a choice of $(d^-(v)-1)$ exits, on the second occasion $(d^-(v)-2)$, and so on. Since the choices at each vertex are independent, there are in all:

$$\prod_{i=1}^{n} (d^-(v_i)-1)!$$

different Eulerian circuits that can be constructed according to the method of theorem 6.4 for a given spanning out-tree. Theorem 6.3 tells us that *every* Eulerian circuit may be associated with a particular spanning out-tree rooted at u. We therefore have the following theorem:

Theorem 6.5. The number of distinct Eulerian circuits in a connected, balanced digraph is given by:

$$T(G)\cdot\prod_{i=1}^{n} (d^-(v_i)-1)!$$

where $T(G)$ is the number of spanning out-trees rooted at a given vertex.

We have already seen how to calculate $T(G)$ for an arbitrary graph in theorem 2.5. For an Eulerian digraph we can draw an immediate conclusion concerning $T(G)$. Since the number of Eulerian circuits cannot depend upon which vertex is taken to be the root in theorem 6.5, it follows that $T(G)$ must also be vertex independent. In other words, in an Eulerian digraph the same number of distinct spanning out-trees are rooted at each vertex.

Figure 6.7 shows an example of counting Eulerian circuits of the digraph G. The diagram shows the Kirchoff matrix $K(G)$. For any r, $1 \leqslant r \leqslant n$, we have $T(G) = \det(K_{rr}(G)) = 2$, whilst $\prod_{i=1}^{n} (d^-(v_i)-1)! = 4$. Hence the number of distinct Eulerian circuits for G is 8. We leave it as a short exercise for the reader to list them.

For a given undirected Eulerian graph G, we could count its Eulerian circuits by noting that it will be the underlying graph of each of a number

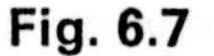

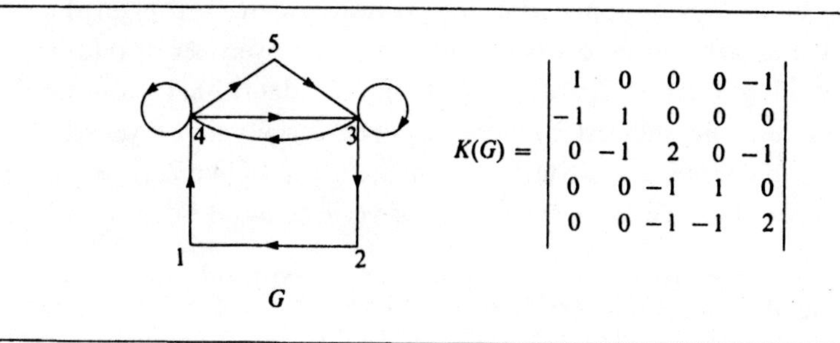

$$K(G) = \begin{vmatrix} 1 & 0 & 0 & 0 & -1 \\ -1 & 1 & 0 & 0 & 0 \\ 0 & -1 & 2 & 0 & -1 \\ 0 & 0 & -1 & 1 & 0 \\ 0 & 0 & -1 & -1 & 2 \end{vmatrix}$$

of determinable Eulerian digraphs. To each Eulerian circuit C of G there will be precisely two corresponding Eulerian circuits in the digraphs, one in, say, G_1 and one in G_2. One of this pair of circuits will correspond to tracing C in one direction in G and the other will correspond to tracing C in the opposite direction. In fact, G_1 will be precisely G_2 with all edge directions reversed.

6.2.2. *The Chinese postman problem for undirected graphs*

We describe here how to find a shortest (non-simple) circuit in a weighted, undirected, non-Eulerian graph such that each edge is traversed at least once. Any postman's circuit, shortest or otherwise, in a non-Eulerian circuit must repeat one or more edges. This is because every vertex is entered the same number of times that it is left, so that any vertex of odd-degree (there must be at least two such vertices) has at least one incident edge that is traversed at least twice. We therefore define $r(u, v)$ to be the number of times that edge (u, v) is repeated in the course of a postman's circuit. In all, (u, v) is traversed $(1+r(u, v))$ times. If we trace a path of repeated edges then we see that it can only end on vertices of odd-degree, perhaps passing through any number of vertices of even-degree (and maybe some of odd-degree) before termination. In any event, the edge repetitions can clearly be partitioned into a set of paths, each path having odd-degree vertices as end-points. Each repetition of an edge belongs to exactly one such path and every vertex of odd-degree is the end-point of just one path. Of course, if we add to the original graph G, $r(u, v)$ repetitions of each edge (u, v) then the resultant graph, which we denote by G'', is Eulerian.

The postman's problem is therefore to find a set of paths such as we have described and such that their edge weight sum is a minimum. The required circuit is then equivalent to an Eulerian circuit of the associated graph G''. A suitable algorithm is described in figure 6.8 and it essentially consists of a series of applications of previously described algorithms. In line 1, the shortest distances between each pair of vertices of odd-degree, in the graph of interest G, are found. A suitable algorithm for this (in fact, it would find the shortest distance between *every* pair of vertices) was described in chapter 1 and verified in theorem 1.5. In line 2, G' is the complete graph whose vertex-set is the set of vertices of odd-degree in G and

Fig. 6.8. Algorithm to solve the Chinese postman problem in an undirected, non-Eulerian graph.

---

1. Find the set of shortest paths between all pairs of vertices of odd-degree in G.
2. Construct G'
3. Find a minimum-weight perfect matching of G'
4. Construct G''
5. Find an Eulerian circuit of G'' and thus a minimum-weight postman's circuit of G.

---

whose edge-weights for each edge (u, v) is $d(u, v)$, the shortest distance from u to v in G. Notice that G' must have an even number of vertices because there are, by theorem 1.1, an even number of vertices of odd-degree in G. The purpose of line 3 is to identify such a matching which has minimum weight. This minimum-weight perfect matching allows us to identify a set of paths of repeated edges (one path from each edge of the matching) needed to solve the Chinese postman problem for G. An efficient minimum-weight perfect matching algorithm is easily contrived from the maximum-weight matching algorithm described in chapter 5. We replace each edge-weight $d(u, v)$ in G' by $(M-d(u, v))$, where M is a constant such that $M > d(u, v)$ for all (u, v). It is then easy to see that a maximum-weight matching in this graph with modified edge-weights is equivalent to a required minimum-weight perfect matching in G'. Line 4 constructs G'' which was defined earlier. Finally, an Eulerian circuit of G'' is found in line 5 (perhaps using the algorithm described in section 6.1.2) which is then easily used to identify a minimum-weight postman's tour of G.

Our description of the algorithm amounts also to its verification. Moreover, notice that the algorithm is efficient because each of its constituent algorithms runs in polynomial time. Figure 6.9 shows an application of the algorithm. The graph G of that diagram is sufficiently simple to identify the $d(u, v)$ by inspection; moreover, we can similarly identify a minimum-weight perfect matching of G' and an Eulerian circuit of G''. The paths of repeated edges in G required for the solution to the Chinese postman problem are (v_1, u, v_4) and (v_2, u_4, v_3).

Fig. 6.9. An example solution to the Chinese postman problem for undirected graphs.

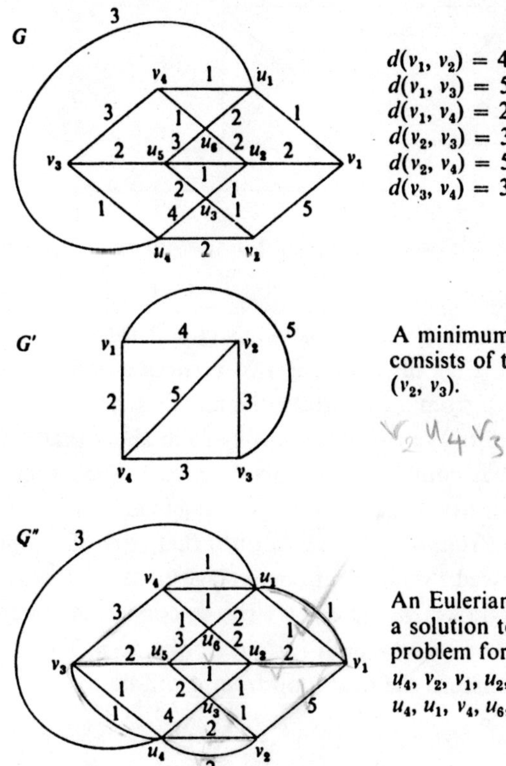

$d(v_1, v_2) = 4$ along (v_1, u_2, u_3, v_2)
$d(v_1, v_3) = 5$ along (v_1, u_2, u_5, v_3)
$d(v_1, v_4) = 2$ along (v_1, u_1, v_4)
$d(v_2, v_3) = 3$ along (v_2, u_4, v_3)
$d(v_2, v_4) = 5$ along $(v_2, u_3, u_2, u_6, v_4)$
$d(v_3, v_4) = 3$ along (v_3, v_4)

A minimum-weight perfect matching consists of the edges (v_1, v_4) and (v_2, v_3).

An Eulerian circuit of G'' and a solution to the Chinese postman problem for G is $(v_1, u_1, v_4, v_3, u_4, v_2, v_1, u_2, u_3, v_2, u_4, u_3, u_5, v_3, u_4, u_1, v_4, u_6, u_5, u_2, u_6, u_1, v_1)$.

6.2.3. *The Chinese postman problem for digraphs*

We consider here directed, weighted graphs. If the graph in question is connected and balanced, then the solution to the Chinese postman problem will be, by corollary 6.1, an Eulerian circuit. Such a circuit may be found by the algorithm of figure 6.5. The remainder of this section describes how to proceed with non-Eulerian digraphs.

In the case of undirected graphs, any connected graph clearly contains a solution to the Chinese postman problem. This is not the case for all connected digraphs. For example, in figure 6.10 no circuit exists which traverses every edge at least once. This is because there is no path from the subset of vertices $\{u_1, u_2, u_3\}$ to the subset $\{v_1, v_2, v_3\}$. The following theorem provides a necessary and sufficient condition for a digraph to contain a postman's circuit.

Fig. 6.10

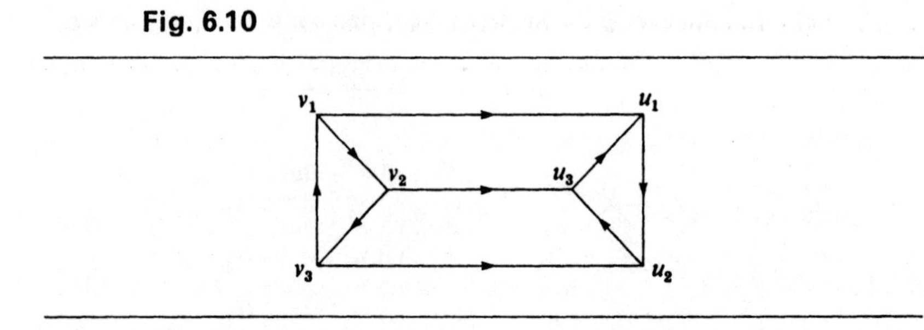

Theorem 6.6. A digraph has a postman's circuit if and only if it is strongly connected.

Proof. Clearly, if the digraph G has a postman's circuit then it must be strongly connected. This is because, for any two vertices u and v part of the circuit provides a path from u to v whilst the remainder of the circuit provides a path from v to u. We therefore only have to show that if G is strongly connected then it contains a postman's circuit. Such a circuit (perhaps a long one) is constructed as follows. Starting from some vertex u we add successive loops (from u and back to u) to that part of the postman's circuit already traced. Suppose at some stage of this process the edge (v_i, v_j) has not been traversed. Since G is strongly connected there will be a path $P(u, v_i)$ from u to v_i and a path $P(v_j, u)$ from v_j to u. To include (v_i, v_j) in the postman's circuit, the next loop from u will be

$$(P(u, v_i), (v_i, v_j), P(v_j, u))$$

We continue until every edge is included. ■

As in the case for undirected graphs, a postman's circuit for a non-Eulerian digraph necessarily involves repeated edges. We again denote the number of times that the edge (u, v) is repeated by $r(u, v)$. Let G'' denote the digraph obtained by adding $r(u, v)$ copies of each edge (u, v) to the original digraph G. Any postman's circuit in G will correspond to an Eulerian circuit of G''. In the case for undirected graphs, the repeated edges formed paths between vertices of odd-degree. In the present case repeated edges must form paths between vertices whose in-degree is not equal to their out-degree. In particular, for any such path from u to v, we must have that:

$$d^-(u) - d^+(u) = D(u) < 0$$

and

$$d^-(v) - d^+(v) = D(v) > 0$$

Moreover, if $D(u) < 0$, then $-D(u)$ paths of repeated edges must start from u. Similarly, if $D(v) > 0$, then $D(v)$ paths must end at v. The problem then reduces to choosing a set of paths such that G'' is balanced and we must do this so as to minimise $\sum_{(u,v)} d(u, v)\, r(u, v)$ where $d(u, v)$ is the weight of (u, v).

The above description suggests the following solution to the Chinese postman problem. It is based upon the flow methodology of chapter 4. Each vertex u, $D(u) > 0$, can be thought of as a source and each vertex v, $D(v) < 0$, can be thought of as a sink. A path of repeated edges from u to v may be thought of as a unit flow with a cost equal to the sum of the edge-weights on the path. In terms of a flow problem, we wish to send (for all u such that $D(u) > 0$), $+D(u)$ units of flow *from* u and (for all v such that $D(v) < 0$), $-D(u)$ units of flow *to* v, and we wish to do this at *minimum cost*. As described in exercise 4.1, we convert this problem of multiple sinks and sources to one of a single source and a single sink. Let the single source be X, then every edge from X to a source u of the original problem is given a capacity equal to $+D(u)$ and is given a cost of zero. Similarly, denoting the single sink by Y, each edge to Y from a sink v of the original problem has a capacity equal to $-D(v)$ and a zero cost. The capacity of all other edges is set to infinity. Since for any digraph:

$$\sum_{u,\, D(u) < 0} D(u) = - \sum_{v,\, D(v) > 0} D(v)$$

a maximum flow (at minimum cost) from X to Y will saturate all edges from X and all edges to Y. Given such a flow we can construct a balanced digraph G''. Any Eulerian circuit in G'' will correspond to a minimum-cost postman's circuit in G.

Fig. 6.11. Algorithm to solve the Chinese postman problem in a non-Eulerian digraph.

1. Construct G'
2. Find a maximum flow at minimum cost in G'
3. Construct G''.
4. Find an Eulerian circuit of G'' and thus a minimum-weight postman's circuit of G.

Given the above description figure 6.11 outlines a suitable algorithm. G' is the network obtained from G by adding the source X and the sink Y, as previously described. Line 2 finds a maximum flow at minimum cost in G'. This can be done by using the minimum-cost flow algorithm of chapter 4. Line 4 might utilise the algorithm of figure 6.5.

As far as the complexity of the algorithm is concerned, notice that the execution time of lines 1, 3 and 4 is bounded by a polynomial in n and $|E|$. In fact, so is the execution time of line 2, but less obviously so. In the previous chapter we saw that the complexity of the minimum-cost flow algorithm is polynomial in n, $|E|$ and the value V of the required flow. The required flow in the present case is given by:

$$V = \sum_{v,\, D(v) > 0} D(v) \leqslant |E|$$

since each edge contributes at most one to the summation. Thus overall, the execution time of this algorithm for the Chinese postman problem for digraphs, is bounded by a polynomial in n and $|E|$ only.

Figure 6.12 shows an application of the algorithm to the graph G of that diagram. G is such that the minimum-cost flow can be found by inspection in G', as can the Eulerian circuit in G''.

Fig. 6.12. An example solution to the Chinese postman problem for digraphs.

G

G'

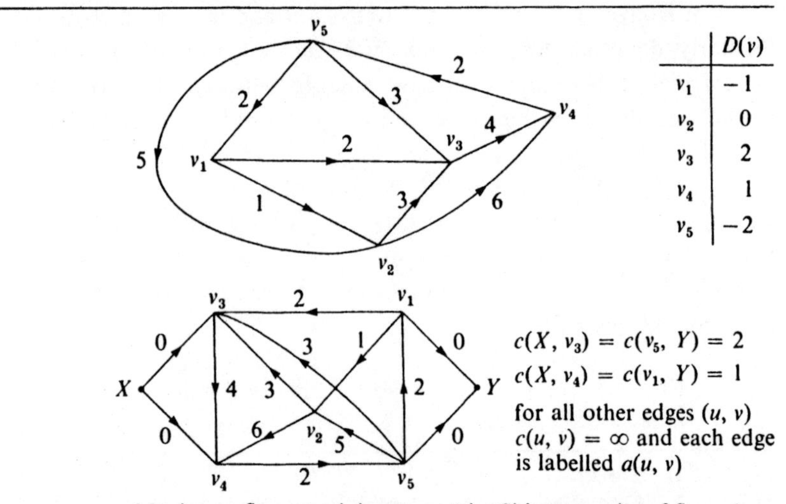

Maximum flow at minimum cost in G' is two units of flow along (X, v_3, v_4, v_5, Y) plus one unit along (X, v_4, v_5, v_1, Y).

G''

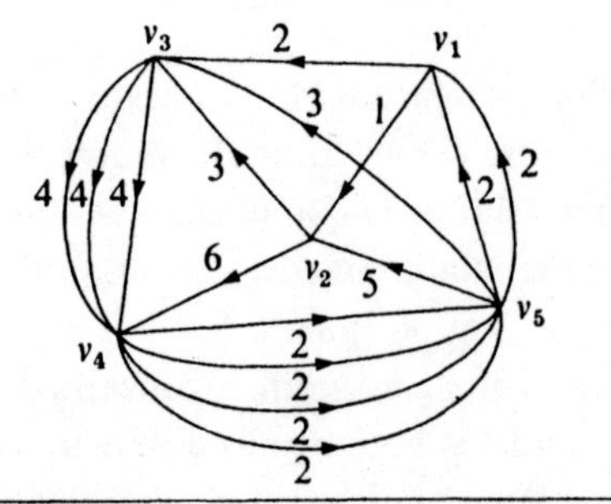

An Eulerian circuit of G'' and a minimum cost postman's circuit of G is $(v_1, v_2, v_3, v_4, v_5, v_2, v_4, v_5, v_3, v_4, v_5, v_1, v_3, v_4, v_5, v_1)$

6.3 Hamiltonian tours

We have already defined a Hamiltonian tour to be a path or a circuit which visits every vertex precisely once. A graph is *Hamiltonian* if it contains a Hamiltonian circuit. Theorem 6.1 provides a quick and simple test for determining whether or not a graph is Eulerian. No such test is known and none is thought to exist to determine whether or not a graph is Hamiltonian. Indeed, the question of whether or not an arbitrary graph is Hamiltonian is a classic *NP*-complete problem. Notice the obvious connection with the problem of finding a longest simple path which we discussed in chapter 1. There are many results which provide either sufficient or necessary conditions for a graph to be Hamiltonian. Section 6.3.1 presents some well-known results in this area.

A well-known problem related to the Hamiltonian circuit problem is that of the *travelling salesman*. The problem is as follows. A salesman, starting in his own city, has to visit each of $(n-1)$ other cities and return home by the shortest route. We shall see that a solution can be provided by finding a Hamiltonian circuit of shortest length in a complete weighted graph. We shall prove in chapter 8 that the question of the existence of a Hamiltonian tour of less than some specified length is *NP*-complete. In section 6.3.2 we describe an inefficient algorithm to find Hamiltonian tours and we also describe some well-known approximation algorithms. These have the advantage of operating in polynomial time but produce results which only approximate, within some known tolerance, to an exact solution.

Many other scheduling problems, see, for example, exercise 6.12, involve consideration of Hamiltonian tours. The origin of interest in these tours is to be found in game theory (exercise 6.3). Puzzles and board games in particular often involve Hamiltonian circuits. For example, the question of finding a *knight's tour* of a chessboard (that is, a sequence of knight's moves which visit every square of a chessboard precisely once and returns to the initial square) is precisely that of finding a Hamiltonian circuit of an associated graph.

6.3.1. Some elementary existence theorems

As we stated earlier, given an arbitrary graph, there is no quick test to determine whether or not it is Hamiltonian. There are, however, many partial results in this area. This section presents some elementary ones.

An immediate observation is that the more edges a graph contains, then the greater is the chance that a Hamiltonian circuit exists. The extreme case is that of complete graphs. Since there is an edge between every pair of vertices and this edge may be traversed in either direction, it follows that any permutation of the vertices is a Hamiltonian path. Moreover, a Hamiltonian circuit is obtained by including the edge from the final to the initial vertex of such a path. We therefore have the following theorem.

Theorem 6.7. Every complete graph is Hamiltonian.

Suppose now that we assign a direction to each edge of a complete graph. The following theorem shows that the resulting digraph still contains a Hamiltonian path.

Theorem 6.8. A digraph, whose underlying graph is complete, contains a Hamiltonian path.

Proof. Let $G = (V, E)$ be a digraph with a complete underlying graph and let $P = (v_1, v_2, \ldots, v_n)$ be a (directed) path in G. Suppose that $v \in V$ is not contained in P. Now, for all i $1 \leqslant i \leqslant n$, we have that:

$$(v, v_i) \notin E \text{ implies that } (v_i, v) \in E$$

and

$$(v_i, v) \notin E \text{ implies that } (v, v_i) \in E$$

Thus v and P may be used to construct a path with $(n+1)$ vertices by the following argument:

If $(v, v_1) \in E$ then the path $(v, v_1, v_2, \ldots, v_n)$ exists,
  otherwise $(v_1, v) \in E$. Then
if $(v, v_2) \in E$ then the path $(v_1, v, v_2, \ldots, v_n)$ exists,
  otherwise $(v_2, v) \in E$. Then
if $(v, v_3) \in E \ldots$
  otherwise $(v_{n-1}, v) \in E$. Then
if $(v, v_n) \in E$ then the path $(v_1, \ldots, v_{n-1}, v, v_n)$ exists,
  otherwise $(v_n, v) \in E$ and
the path $(v_1, v_2, \ldots, v_n, v)$ exists.

Hence, starting with any path (a single edge would do) we can repeatedly extend it by the addition of vertices until every vertex is included. ■

It is easy to construct examples to show that a digraph satisfying the last theorem need not be Hamiltonian. However, the next theorem provides a narrower definition which guarantees the presence of a Hamiltonian circuit.

Theorem 6.9. A strongly connected digraph whose underlying graph is complete is Hamiltonian.

Proof. If $G = (V, E)$ satisfies the theorem then being strongly connected it contains at least one simple circuit. Let $C = (v_1, v_2, \ldots, v_n, v_1)$ be such a circuit of maximum length. Suppose that C does not include some vertex v. We first show that *either* there is an edge incident from v to every vertex v_i of C *or* there is an edge incident from every vertext of C to v. Since C is maximal, then, for all i, modular n:

$$(v, v_i) \in E \text{ implies that } (v_{i-1}, v) \notin E$$

and

$$(v_i, v) \in E \text{ implies that } (v, v_{i+1}) \notin E$$

otherwise, in either case a circuit longer than C could be constructed. Now since G has an underlying complete graph:

$$(v_{i-1}, v) \notin E \text{ implies that } (v, v_{i-1}) \in E$$

and

$$(v, v_{i+1}) \notin E \text{ implies that } (v_{i+1}, v) \in E$$

Hence for all i:

$$(v, v_i) \in E \text{ implies that } (v, v_{i-1}) \in E$$

and

$$(v_i, v) \in E \text{ implies that } (v_{i+1}, v) \in E$$

We can therefore partition those vertices not on C into two classes, V' and V''. There is an edge incident from each vertex in V' to every vertex of C, and there is an edge incident from each vertex in C to every vertex in V''.

Now, since our hypothesis is that C is not a Hamiltonian circuit, we have that $V' \cup V'' \neq \varnothing$. Moreover, G is strongly connected so that $V' \neq \varnothing$ *and* $V'' \neq \varnothing$, and there exists an edge from V'' to V'. Denoting this edge by (v'', v'), we then have a circuit $C' = (v_1, v_2, \ldots, v_n, v'', v', v_1)$ such that $|C'| > |C|$. Therefore our hypothesis is contradicted and C must be a Hamiltonian circuit of G. ■

Let us return to undirected graphs. Theorem 6.7 is hardly a powerful theorem. The degree of any vertex in a complete graph is $(n-1)$ and this is also the minimum degree δ, of any vertex in that graph. Theorem 6.10 provides a rather stronger result in the sense of guaranteeing a Hamiltonian circuit for a smaller value of δ.

Theorem 6.10. If G is a graph such that $n > 3$ and $\delta > \frac{1}{2}n$ then G is Hamiltonian.

Proof. Suppose that G satisfies the conditions of the theorem but that it is not Hamiltonian. G cannot therefore be complete because of theorem 6.7. We can add edges to G without violating the conditions of the theorem until the addition of any one extra edge will create a Hamiltonian circuit. Let v_1 and v_n now be any two non-adjacent vertices of G. Now $G+(v_1, v_n)$

is Hamiltonian so that G contains a Hamiltonian path, $(v_1, v_2, v_3, \ldots, v_n)$, from v_1 to v_n. We now define two subsets of vertices:

$$V' = \{v_i | (v_1, v_{i+1}) \in E\}$$

and

$$V'' = \{v_i | (v_i, v_n) \in E\}$$

Now $|V' \cap V''| = 0$. If this were not the case then V' and V'' would contain a common vertex v_i and G would contain the Hamiltonian circuit $(v_1, v_2, \ldots, v_i, v_n, v_{n-1}, \ldots, v_{i+1}, v_1)$. Also, $|V' \cup V''| < n$ because v_n is in neither V' nor V''. We therefore see that:

$$d(v_1) + d(v_n) = |V'| + |V''| = |V' \cup V''| + |V' \cap V''| < n$$

so that even with the additional edges added to G we have that

$$d(v_1) \text{ or } d(v_n) < \tfrac{1}{2}n$$

This contradicts the original assumption so that G must be Hamiltonian. ■

The above theorems provide *sufficient* conditions for the existence of Hamiltonian tours. It is a trivial matter to construct examples showing that these conditions are not *necessary*. Our final theorem provides a necessary condition for a graph, or indeed a digraph, to be Hamiltonian.

Theorem 6.11. If $G = (V, E)$ is Hamiltonian, then for every non-empty proper subset of vertices $V' \subset V$:

$$C(G - V') \leqslant |V'|$$

where $C(G - V')$ is the number of components of the graph $(G - V')$.

Proof. If H is a Hamiltonian cycle of G and therefore a spanning subgraph of G, we have for every V':

$$C(H - V') \leqslant |V'|$$

But $(H - V')$ is a spanning subgraph of $(G - V')$ and so:

$$C(G - V') \leqslant C(H - V')$$

and so the theorem follows. ■

As an example of the use of theorem 6.11 consider the graph of figure 6.13(*a*). The removal of the vertices v_1 and v_2 leaves the three components shown in figure 6.13(*b*). Hence for this graph G, we have that $C(G - \{v_1, v_2\}) > |\{v_1, v_2\}|$, and so by theorem 6.11 G cannot be Hamiltonian.

The condition of theorem 6.11 is not sufficient for a graph to be Hamiltonian. For example, the Petersen graph shown in figure 6.14 is not Hamiltonian (see exercise 6.8) and yet every non-empty proper subset of its vertices, V', satisfies $|V'| \geqslant C(G - V')$.

Fig. 6.13

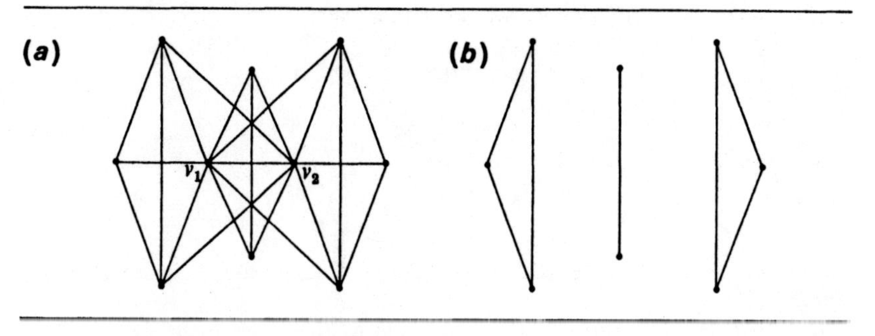

Fig. 6.14. The Petersen graph.

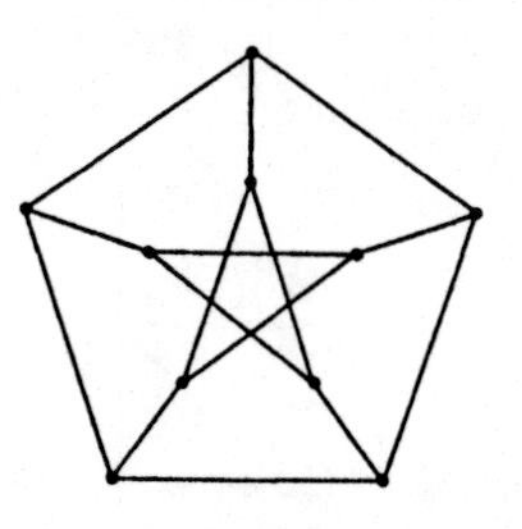

We have described here just a few of the most elementary theorems concerning the existence of Hamiltonian tours. There are many other results in this area. See, for example, chapter 10 of Berge[2].

6.3.2. Finding all Hamiltonian tours by matricial products

We describe here, by example, a straightforward technique for generating all the Hamiltonian tours of a graph or digraph. It can easily be adapted to find a shortest Hamiltonian circuit. Like all known, and probably unknown, methods for this problem, it provides an inefficient solution.

In chapter 1 we saw how the (i, j)th element of the kth power of the adjacency matrix of a graph gives the number of paths of length k from vertex i to vertex j. That is, the number of complex paths. We now see a variation of this theme where an element of a related matricial product individually identifies each simple path of length k from i to j. Such a path of length $(n-1)$, where n is the number of vertices, is necessarily a Hamiltonian path. Given all the Hamiltonian paths it is a trivial matter to identify all the Hamiltonian circuits.

Fig. 6.15

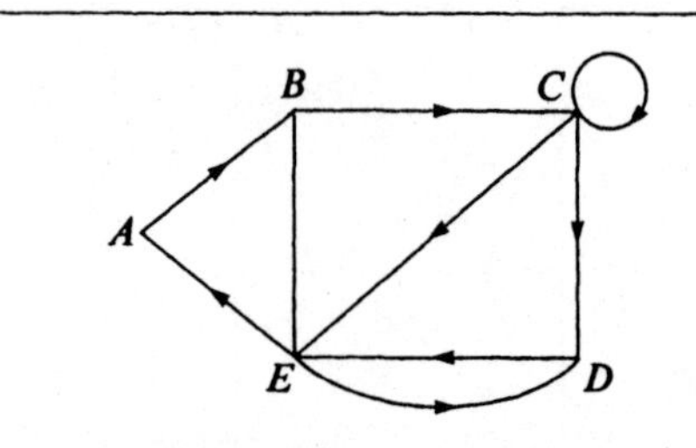

We illustrate the method with reference to the digraph of figure 6.15. Application to an undirected graph should be obvious. First we construct a matrix M_1, formed from the adjacency matrix by replacing any (i, j)th non-zero entry with the string ij, and any non-zero diagonal element is replaced by zero. For our example:

$$M_1 = \begin{vmatrix} 0 & AB & 0 & 0 & 0 \\ 0 & 0 & BC & 0 & 0 \\ 0 & 0 & 0 & CD & CE \\ 0 & 0 & 0 & 0 & DE \\ EA & EB & 0 & ED & 0 \end{vmatrix}$$

We now define a second matrix, M, derived from M_1 by deleting the initial letter in each element that is a string, for our example:

$$M = \begin{vmatrix} 0 & B & 0 & 0 & 0 \\ 0 & 0 & C & 0 & 0 \\ 0 & 0 & 0 & D & E \\ 0 & 0 & 0 & 0 & E \\ A & B & 0 & D & 0 \end{vmatrix}$$

Finally, we define a matricial product from which we can generate M_j, for all j where $n > j > 1$. M_j displays all the simple paths of length j:

$$M_j = M_{j-1} * M$$

where the (r, s)th element of M_j is defined as follows:

$$M_j(r, s) = \left\{ \hat{M}_{j-1}(r, t)\, M(t, s) \;\middle|\; \begin{matrix} 1 \leqslant t \leqslant n,\ \hat{M}_{j-1}(r, t) \in M_{j-1}(r, t): \\ \text{neither } \hat{M}_{j-1}(r, t) \text{ nor } M(t, s) \text{ are} \\ \text{zero or have a common vertex} \end{matrix} \right\}$$

Here $\hat{M}_j(r, t)M(t, s)$ denotes the concatenation of $\hat{M}_j(r, t)$ and $M(t, s)$. Clearly, $M_j(r, s)$ is the set of simple paths (since by construction no vertex appears more than once in any path) from r to s consisting of j edges. Using this definition in our example we obtain:

$$M_2 = \begin{vmatrix} 0 & 0 & ABC & 0 & 0 \\ 0 & 0 & 0 & BCD & BCE \\ CEA & CEB & 0 & CED & CDE \\ DEA & DEB & 0 & 0 & 0 \\ 0 & EAB & EBC & 0 & 0 \end{vmatrix}$$

$$M_3 = \begin{vmatrix} 0 & 0 & 0 & ABCD & ABCE \\ BCEA & 0 & 0 & BCED & BCDE \\ CDEA & \begin{pmatrix} CEAB \\ CDEB \end{pmatrix} & 0 & 0 & 0 \\ 0 & DEAB & DEBC & 0 & 0 \\ 0 & 0 & EABC & EABD & 0 \end{vmatrix}$$

$$M_4 = \begin{vmatrix} 0 & 0 & 0 & ABCED & ABCDE \\ BCDEA & 0 & 0 & 0 & 0 \\ 0 & CDEAB & 0 & 0 & 0 \\ 0 & 0 & DEABC & 0 & 0 \\ 0 & 0 & 0 & EABCD & 0 \end{vmatrix}$$

Generally each matricial element is a set of paths, although the only entry in our example which consists of more than one path is $M_3(C, B)$. Since our example has $n = 5$, M_4 displays all the Hamiltonian path of the graph of figure 6.15. In order to establish the Hamiltonian cycles we need only check whether or not the end-points of these six paths are appropriately connected by an edge. This establishes a single Hamiltonian circuit, namely (A, B, C, D, E, A). Alternatively, we could carry out a final multiplication, $M_n = M_{n-1} * M$, for which the requirement that no vertex appears more than once is dropped. The diagonal entries in M_n would describe any Hamiltonian circuits.

There are many ways of finding Hamiltonian tours of a graph (if they exist). The one described here is particularly straightforward which is in contrast to other methods, for example the branch and bound algorithm, more commonly encountered. All of these methods are inefficient involving unacceptable volumes of computation. This is because, as we indicate in the next section, the number of potential circuits grows factorially with n.

6.3.3. The travelling salesman problem

There are several extant definitions of the travelling salesman problem. One definition requires us to find the minimum-length circuit which visits every vertex of a weighted graph *precisely once*. Another definition requires us to find a minimum-length circuit of a weighted graph which visits every vertex *at least once.* For the following discussion we take the second definition. Figure 6.16 illustrates that a solution to this travelling salesman problem (in our example the circuit (a, b, a, c, a) with length 4) is not necessarily a simple circuit. This is generally true for any

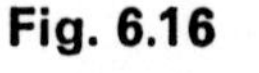

Fig. 6.16

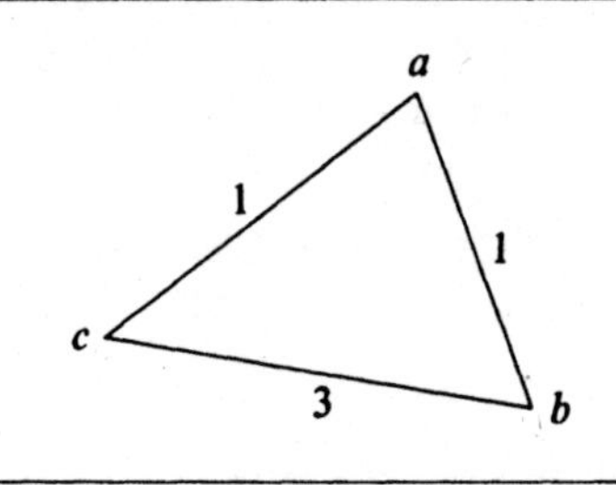

graph in which the *triangle inequality* does not hold. If for every pair of vertices u and v of a graph G, the weight $w(u, v)$ satisfies

$$w(u, v) \leqslant w(u, x)+w(x, v)$$

for all vertices $x \neq u$, $x \neq v$, then the triangle inequality is said to be satisfied in G. Notice that if G is, say, a representation of a road network and the edge-weights are actual distances, then the triangle inequality will almost certainly hold in G. However, if the edge-weights represent some other quantity, say, for example, the cost of transportation, then it could well be that the triangle inequality is not satisfied. It is useful for such cases to notice that there is a simple technique for converting the travelling salesman problem for any graph $G = (V, E)$ into the problem of finding a minimum-weight Hamiltonian circuit for another graph $G' = (V', E')$. G' is a complete graph with $V = V'$ and each edge $(u, v) \in E'$ has a weight $w(u, v)$ equal to the minimum distance from u to v in G. Notice that each edge of G' corresponds to a path of one or more edges in G. In constructing G' it is useful to label any edge with the path it represents in G if this path is longer than one edge. Figure 6.17 shows G' for G of figure 6.16. Given G and G' as just defined we have the following theorem. We re-emphasise that if triangle inequality does not hold, then the travelling salesman problem means a shortest circuit visiting each vertex *at least* once and not *precisely* once. If triangle inequality is satisfied, then either definition will do.

Theorem 6.12. A solution to the travelling salesman problem in G corresponds to and is the same length as a minimum-weight Hamiltonian circuit in the complete graph G'.

Fig. 6.17

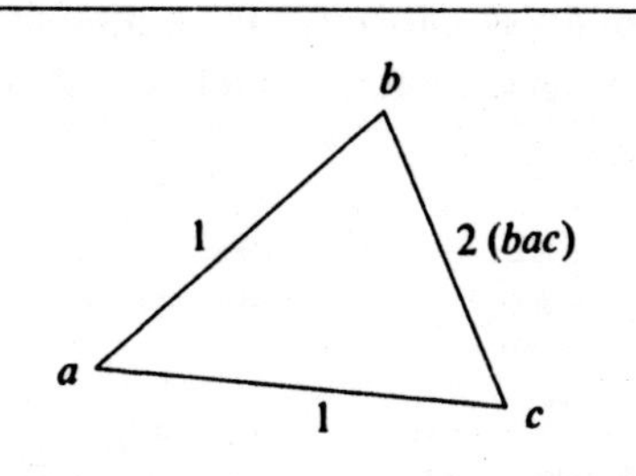

Proof. Suppose that C is a solution to the travelling salesman problem for G which is *not* equivalent to a minimum-weight Hamiltonian circuit of G'. Let C' be the equivalent circuit in G'. Notice that C' must follow the same sequence of edges in G' as C does in G and that these edges have the same weight in G' as in G. This is because if any one of these edges, say (u, v), had a smaller weight in G', then a shorter solution to the travelling salesman problem could be found by replacing (u, v) in C by the sequence of edges that labels (u, v) in G'. C' is a circuit in G' which visits every vertex at least once.

Suppose that C' visits some vertex s a second time. Let r be the vertex visited just before this happens and let t be the vertex visited just after it happens. We can replace the subpath (r, s, t) in C' by (r, t) without affecting the fact that C' is equivalent to a solution to the travelling salesman problem in G. This is because, by construction $w(s, t)$ in G' is equal to the length of the shortest path from s to t in G. In this way we can eliminate any multiple visitations to any vertex in G' and, contrary to our hypothesis, C' eventually becomes a Hamiltonian circuit. Notice that it must be a Hamiltonian circuit of minimum length; a Hamiltonian circuit of minimum length in G' would be equivalent to a shorter solution to the travelling salesman problem in G. ■

In view of the previous theorem we can assume from now on that we wish, in solving the travelling salesman problem, to find a minimum-weight Hamiltonian circuit in a complete graph. This is unless, of course, we mean, by the travelling salesman problem, a circuit of shortest length which visits every vertex *precisely* once in a graph for which triangle inequality does not hold. An immediately obvious method of solution is to enumerate all the Hamiltonian circuits and then by comparison to find the shortest. This approach, although straightforward, presents us with an unacceptably large amount of computation. For a complete undirected graph with n vertices, there are $\frac{1}{2}(n-1)!$ essentially different Hamiltonian circuits. The number of addition operations then required to find the lengths of all these circuits is $O(n!)$. Given a computer that can perform these additions at the rate of 10^8/second, the following approximate computation times follow:

n	$\sim n!$	Time
12	4.8×10^8	5 seconds
15	1.3×10^{12}	3 hours
20	2.4×10^{18}	800 years
50	3.0×10^{64}	10^{49} years

The importance of this model calculation is that it demonstrates the phenomenal rate of growth of the computation time as n increases. For n of quite moderate value, $n!$ is too large to make the computation feasible. In fact, no known efficient algorithm exists for the travelling salesman problem. In chapter 8, we prove that the problem of determining whether or not a Hamiltonian circuit exists, which is shorter than a specified length, is *NP*-complete.

For the travelling salesman problem, as indeed for any other intractable problem, it is useful to have a polynomial time algorithm which will produce, within known bounds, an approximation to the required result. Such algorithms are called *approximation algorithms*. Let L be the value obtained (for example, this might be the length of a travelling salesman's circuit) by an approximation algorithm and let L_0 be an exact value. We require a performance guarantee for the approximation algorithm which could, for a minimisation problem, be stated in the form:

$$1 \leqslant L/L_0 \leqslant \alpha$$

For a maximisation problem we invert the ratio L/L_0. Of course, we would like α to be as close to one as possible.

Unfortunately, not every heuristic produces a useful approximation algorithm. Consider the following approach which is perhaps the most immediately obvious for the travelling salesman. Starting at vertex v_1, we trace C, an approximation to a minimum-weight Hamiltonian circuit, along (v_1, v_2) which is the shortest edge from v_1. We then leave v_2 along (v_2, v_3), the shortest edge from v_2 which keeps C acyclic. We continue in this way until every vertex has been visited. The Hamiltonian circuit is then completed by the edge (v_n, v_1). It can be shown (see Liu[11]) that for this algorithm:

$$\alpha = \tfrac{1}{2}(\lceil \ln n \rceil + 1)$$

Thus, for this so-called *nearest-neighbour* method, the possible error in the approximation is a function of the problem size. For arbitrarily large graphs, the resulting error may be arbitrarily large. Fortunately, we can do better than this.

Consider the approximation algorithm presented in figure 6.18. G, which is subjected to the algorithm, is a complete weighted graph within which the triangle inequality holds. The algorithm first finds a minimum-weight spanning-tree T of G using, perhaps, Prim's algorithm described in chapter 2. The next step associates a depth-first index to each vertex with respect to a depth-first search of T. For this, the depth-first algorithm described in chapter 1 could be used. Finally, the algorithm outputs a Hamiltonian circuit which visits the vertices of G in the order of the depth-first indices.

Fig. 6.18. An approximation algorithm for the travelling salesman problem.

---

1. Find a minimum-weight spanning-tree T of G.
2. Conduct a depth-first search of T associating a depth-first index $L(v)$ with each vertex v.
3. Output the following approximate minimum-weight Hamiltonian circuit:

$$C = (v_{i_1}, v_{i_2}, v_{i_3}, \ldots, v_{i_n}, v_{i_1})$$

where $L(v_{i_j}) = j$

The component steps of this algorithm, as described in previous chapters, have low order polynomial time complexity and so the algorithm is an efficient one.

Figure 6.19 shows an application of this algorithm to the graph G. A minimum-weight spanning-tree T consists of the heavily scored edges. C is the Hamiltonian circuit output by the algorithm. The subscript on each vertex denotes its depth-first index with respect to the particular traversal of T undertaken. C_0 indicated in the diagram is a circuit which, in a depth-first traversal of T, travels twice around the spanning-tree T. Here, C_0, closely related to C, gives the algorithm its name of the *twice-around-the-minimum-weight-spanning-tree* algorithm.

Theorem 6.13. For any travelling salesman problem within which the triangle inequality is satisfied, the twice-around-the-minimum-weight-spanning-tree algorithm gives $\alpha < 2$.

Proof. Let W be the weight of a minimum-weight spanning-tree of G and let W_0 be the weight of a minimum-weight Hamiltonian circuit of G. We first note that:

$$W < W_0$$

because a spanning-tree shorter than a minimum-length Hamiltonian circuit can be obtained by deleting any edge from this circuit. We next observe that a depth-first search of a spanning-tree traces a circuit C_0 which traverses each edge of the tree twice. For a minimum-weight tree this circuit has length $2W$ which is strictly less than $2W_0$. Now the circuit C generated by the algorithim follows C_0 except that C proceeds directly to the next unvisited vertex on C_0 rather than revisiting any vertices. Because the triangle inequality holds within G, C is no longer than C_0 and so the theorem follows. ■

Fig. 6.19. An application of the twice-around-the-minimum-weight-spanning-tree algorithm.

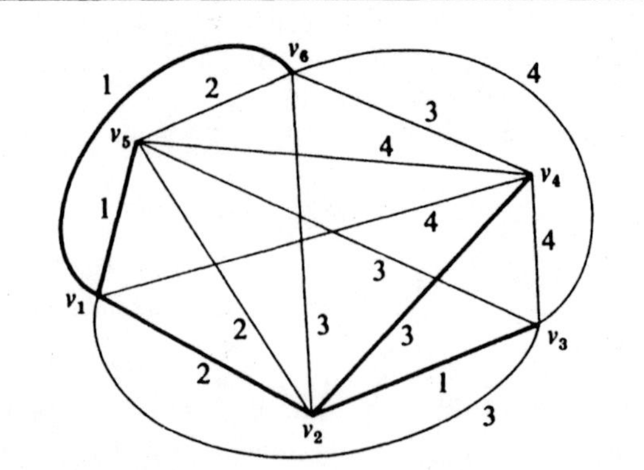

$$C_0 = (v_1, v_2, v_3, v_2, v_4, v_2, v_1, v_5, v_1, v_6, v_1),\ C = (v_1, v_2, v_3, v_4, v_5, v_6, v_1)$$

The heuristics used to obtain a circuit from a minimum-weight spanning-tree in the above algorithm can be improved upon as we describe in figure 6.20. Here steps 2, 3 and 4 essentially replace step 2 of the previous algorithm. Notice that in step 2 of figure 6.20, V' must contain an even number of vertices by theorem 1.1. Thus a perfect matching of V' exists and one of minimum weight can be found in polynomial time in just the same way as was described for step 3 of the algorithm of figure 6.8. In step 3 of figure 6.20, G' must be Eulerian because the construction ensures that every vertex is of even-degree. Step 4 may be efficiently carried out incorporating one of the Eulerian circuit algorithms described elsewhere in this chapter. For convenience we shall call the algorithm of figure 6.20, the minimum-weight matching algorithm for the travelling salesman problem.

Fig. 6.20. An improved approximation algorithm for the travelling salesman problem.

1. Find a minimum-weight spanning-tree T of G.
2. Construct the set V' of vertices of odd-degree in T and find a minimum-weight perfect matching M for V'.
3. Construct the Eulerian graph G' obtained by adding the edges of M to T.
4. Find an Eulerian circuit C_0 of G' and index each vertex according to the order, $L(v)$, in which v is first visited in a trace of C_0.
5. Output the following approximate minimum-weight Hamiltonian circuit:

   $C = (v_{i_1}, v_{i_2}, v_{i_3}, \ldots, v_{i_n}, v_{i_1})$

   where $L(v_{i_j}) = j$

Figure 6.21 shows an application of this algorithm to the graph G of figure 6.19. In figure 6.21 a minimum-weight tree T of G is indicated by heavily scored edges. In this case every vertex of T has odd-degree and so belongs to V'. A minimum-weight perfect matching of V' is, by inspection, $M = \{(v_1, v_5), (v_2, v_3), (v_4, v_6)\}$. G', constructed from the edges of M and T, is shown in the diagram. C_0 is the Eulerian circuit from which the tabulated indices $L(v)$ are derived. From these the approximate minimum-length Hamiltonian circuit C is obtained.

Fig. 6.21. An application of the minimum-weight matching algorithm for the travelling salesman problem.

$C_0 = (v_1, v_5, v_1, v_2, v_3, v_2, v_4, v_6, v_1)$

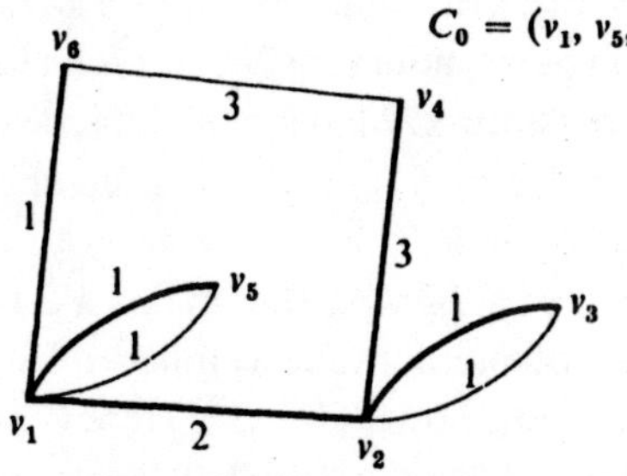

v	$L(v)$
v_1	1
v_2	3
v_3	4
v_4	5
v_5	2
v_6	6

$C = (v_1, v_5, v_2, v_3, v_4, v_6, v_1)$

The following theorem shows that the present heuristics are an improvement on those used in the algorithm of figure 6.18.

Theorem 6.14. For any travelling salesman problem within which the triangle inequality is satisfied, the minimum-weight matching algorithm for the travelling salesman problem gives $\alpha < \frac{3}{2}$.

Proof. Because the triangle inequality holds, the circuit C is no longer than the circuit C_0. As in the previous algorithm C follows C_0 except that C proceeds directly to the next unvisited vertex of C_0 rather than revisiting any vertices. The weight of C_0 is obtained by adding the weight of T, which we denote by W, to the weight of the matching M, which we denote by W_1. By W_0 we denote the weight of a minimum-weight Hamiltonian circuit of G. As in theorem 6.13:

$$W < W_0$$

and we just need to show that $W_1 \leqslant \frac{1}{2}W_0$ in order to complete the proof.

Given a Hamiltonian circuit H of weight W_0, we can construct a circuit of no greater weight which passes only through the vertices in V'. We do this by tracing H and by-passing those vertices not in V'. Because the triangle inequality holds, the new circuit cannot be longer than H. Because V' contains an even number of vertices we can construct two matchings from this new circuit, each obtained by taking alternate edges. Consider that matching of this pair which has smallest weight. This matching has a weight which is not less than W_1 but which is not greater than half the weight of the circuit through the vertices of V'. The result follows. ∎

At present no polynomial time algorithm is known which gives a better approximation guarantee than that provided by theorem 6.14.

6.3.4. 2-factors of a graph

We define a *k-factor* of a graph G to be a k-regular spanning subgraph of G. Our interest here concerns 2-factors because a Hamiltonian circuit is a 2-factor, although, of course, not every 2-factor is a Hamiltonian circuit. For example, the graph of figure 6.22 has several 2-factors including a Hamiltonian circuit (1, 2, 3, 4, 7, 8, 6, 5, 1) and the 2-factor with component circuits (1, 2, 6, 5, 1) and (3, 4, 8, 7, 3).

We can determine whether or not a digraph G contains a 2-factor, and find an example if one exists, in polynomial time as follows. If $G = (V, E)$ then we first construct a bipartite graph $G' = (V'', E')$. Here $V'' = V \cup V'$, $|V| = |V'|$ and G' has the bipartition (V, V'). We shall denote the vertices in V by $v_1, v_2, \ldots, v_n$ and those in V' by $v'_1, v'_2, \ldots, v'_n$. There is an edge $(v_i, v'_j) \in E'$ if and only if $(v_i, v_j) \in E$. The construction is illustrated for

Fig. 6.22

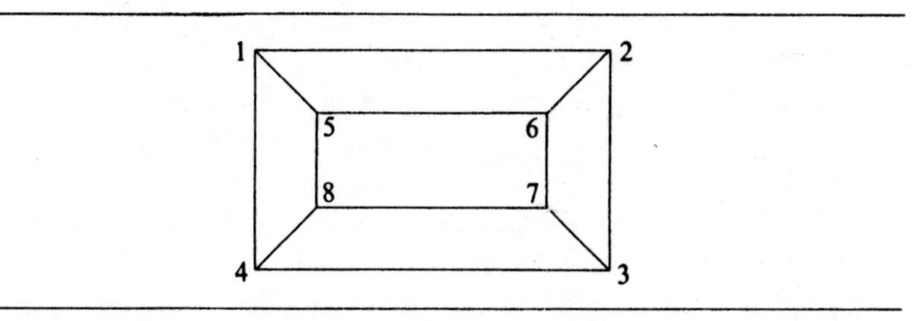

G shown in figure 6.23. The second step of the algorithm is to find a maximum matching in the graph G'. This can be done in polynomial time, as described at the beginning of section 5.2. Such a matching is indicated by heavily scored edges in figure 6.23. We now need the following theorem.

Fig. 6.23

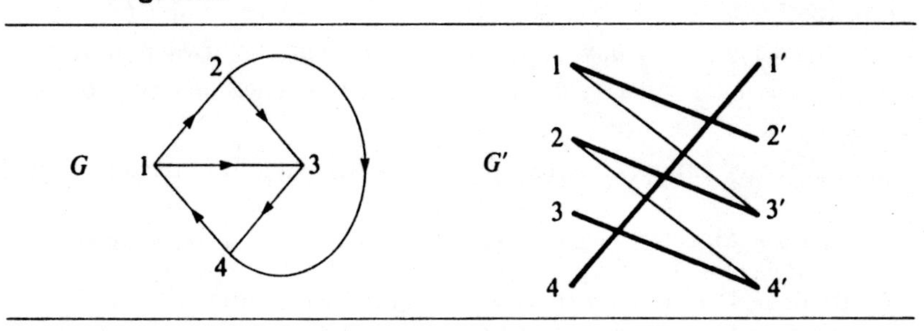

Theorem 6.15. G contains a 2-factor if and only if G' contains a perfect matching.

Proof. Suppose that G' contains a perfect matching M. It is easy to see that G then contains a 2-factor which consists of every edge (v_i, v_j) such that $(v_i, v_j') \in M$. Conversely, suppose that G contains a 2-factor. If (v_i, v_j) is an edge of the 2-factor, then (v_i, v_j') is an edge of a matching in G'. It is easy to see that this matching must be perfect. ■

If the maximum matching phase of our 2-factor algorithm finds a matching M, such that $|M| = n$, then M will be perfect. In this case, according to theorem 6.15, G contains a 2-factor which consists of every edge (v_i, v_j) such that $(v_i, v_j') \in M$.

6.4 Summary and references

We can determine the existence of and find an Eulerian circuit of a graph or digraph in linear-time. This is in marked contrast to the situation for Hamiltonian circuits. As we shall prove in chapter 8, the question of whether or not a graph is Hamiltonian is *NP*-complete. Of the existence theorems of section 6.3.1, theorem 6.10 is due to Dirac.[3] Many others can be found in the expository accounts in chapter 10 of Berge[2] and chapter 6 of Beineke & Wilson.[4]

Scheduling problems often involve Eulerian and Hamiltonian circuits. Edmonds & Johnson[5] provide a comprehensive treatment of postman problems and methods of Eulerian circuit generation. There is a great volume of literature associated with the travelling salesman problem. See, for example, the survey of Bellmore & Nemhauser.[6] The minimum-weight matching approximation algorithm described in the text is due to Christofides.[7] Although we were able to prove the effectiveness of such an algorithm for one class of travelling salesman problems, Sahni & Gonzalez[8] have shown that the problem is non-approximable (unless the *NP*-complete problems have polynomial time solutions) if we require both that the salesman must visit each city precisely once *and* that triangle inequality is not satisfied.

For general reading, Chapters 6 and 7 of Mineaka[9] are recommended.

[1] Kuan, M-K. 'Graphic programming using odd or even points', *Chinese Math.*, **1**, 273–7 (1962).
[2] Berge, C. *Graphs and Hypergraphs.* North-Holland (1973).
[3] Dirac, G. A. 'Some theorems on abstract graphs', *Proc. London Maths Soc.* **2**, 69–81 (1952).
[4] Beineke, L. W. & Wilson, R. J. (eds), *Selected Topics in Graph Theory*, chapter 6 (J. C. Bermond). Academic Press (1978).
[5] Edmonds, J. & Johnson, Ellis, L. 'Matching, Euler tours and the Chinese postman', *Math. Prog.*, **5**, 88–124 (1973).
[6] Bellmore, M. & Nemhauser, G. L. 'The travelling salesman problem: a survey', *Operations Research*, **16**, 538–58 (1968).
[7] Christofides, N. 'Worst-case analysis of a new heuristic for the travelling salesman problem', *Technical Report*, Graduate School of Industrial Administration, Carnegie–Melon University, Pittsburgh, PA (1976).
[8] Sahni, S. & Gonzalez, T. '*P*-complete approximation problems', *JACM*, **23**, 555–65 (1976).
[9] Mineaka, E. *Optimisation Algorithms for Networks and Graphs.* Marcel Dekker (1978).
[10] Even, S. *Graph Algorithms.* Computer Science Press (1979).
[11] Liu, C. L. *Elements of Discrete Mathematics*, chapter 4. McGraw-Hill, Computer Science Series (1977).

EXERCISES

6.1. In what graphs is an Eulerian circuit also a Hamiltonian circuit?

6.2. In 1736 (theorem 6.1) Euler solved a recreational puzzle of interest to the inhabitants of Königsberg (now Kaliningrad). Kaliningrad sits across the river Pregel with seven bridges connecting the various banks and islands of the river as shown. The problem is whether or not it is possible to follow a circular walk starting and finishing at the same river bank and crossing each bridge precisely once. What is the answer?

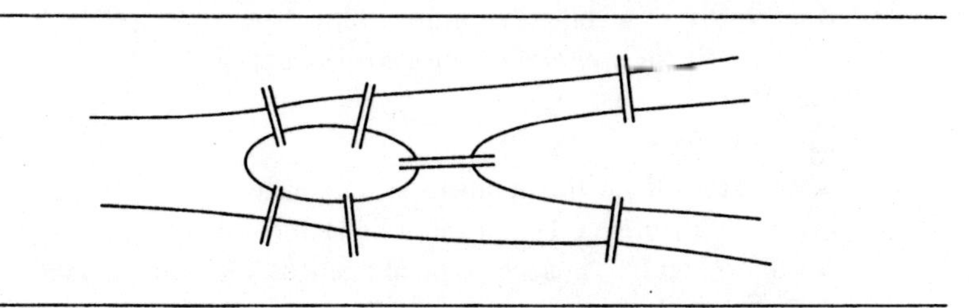

6.3. In 1859 Sir William Hamilton sold, for 25 guineas, a puzzle to a Dublin games manufacturer. The puzzle consisted of a dodecahedron (a regular solid figure with 12 pentagonal faces and hence 20 corners) and on each corner was marked the name of some capital city. One game that could be played was to construct a world tour. This consisted of a circuit, following the edges of the dodecahedron, which visited every capital city exactly once. Trace such a Hamiltonian circuit on the projection of the dodecahedron below.

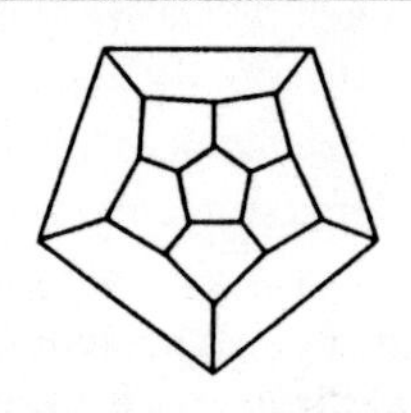

6.4. A *tournament* is a digraph in which there is precisely *one* directed edge between any pair of vertices. Suppose that n people play in a singles tennis competition, each player meeting each of the other $(n-1)$ competitors just once. Clearly, a tournament is a representation of the competition results in which the edge (i, j) implies that competitor i beat competitor j. Show that the competitors can always be ordered so that any competitor is immediately above a competitor he has beaten (see theorem 6.8). In general how quickly can such an ordering be found?

Suppose that for an edge (i, j) there also exists a path from j to i. Comment upon the sensibility of ranking the players now in the way described.

6.5. Using the proof of theorem 6.1 as a basis, carefully describe the details of an algorithm of $O(|E|)$ complexity which finds an Eulerian circuit of an undirected Eulerian graph.

6.6. Determine the complexity of the algorithm described in section 6.3.2 to find all the Hamiltonian tours of a graph. Hence justify the claim that it is inefficient.

6.7. $G = (V, E)$ is a bipartite graph with bipartition (V', V''), where $|V'| \neq |V''|$. Show that there always exists a proper subset of vertices W such that:

$C(G-W) > W$

where $C(G-W)$ is the number of components of $(G-W)$. Therefore, in view of theorem 6.11, G cannot be Hamiltonian.

6.8. (*a*) Show that the Petersen graph of figure 6.14 is not Hamiltonian. The amount of computation required to find all the Hamiltonian circuits of a graph with ten vertices using the usual algorithms will be large, so use *ad hoc* arguments.

(*b*) Demonstrate that the removal of any vertex from the Petersen graph yields a Hamiltonian graph. (The Petersen graph is the only non-Hamiltonian graph with ten or less vertices with this property.)

6.9. Show that every 3-regular graph without cut-edges, contains a 2-factor (see exercise 5.6). Notice (exercise 6.8(*a*)) that not every such graph is Hamiltonian.

6.10. The Chinese postman problem for both non-Eulerian graphs and non-Eulerian digraphs has, as indicated in the text, an efficient solution. Obtain polynomial bounds, which are as tight as you can make them, for the execution times of the algorithms described.

6.11. The following problem may appear in a number of guises. In essence it amounts to finding a longest circular sequence of characters (from an alphabet of m characters $l_1, l_2, \ldots, l_m$) that can be formed without repeating a subsequence of n characters. Such a sequence is called a *de Bruijn* sequence and for $m = 2$ the problem is called the *teleprinters problem*. Since there are m^n distinct subsequences, then the required sequence cannot be more than m^n characters long. Does a sequence of this length exist and if so, how can it be constructed? The problem can be solved using a graph in which each edge is labelled with one of the n-character subsequences; if one edge follows another on some path of the graph then the construction is such that their labels are of possible contiguous subsequences in the required circular sequence. For example, if we denote a particular $(n-1)$ sequence of characters by α and if $l_i\alpha$ labels an edge into some vertex, then those edges labelled $\alpha l_1, \alpha l_2, \ldots, \alpha l_m$ will leave that vertex. In the diagram below, (*a*) shows the general attachment of edges to any vertex which is naturally labelled α; (*b*) shows the whole graph for $n = m = 3$.

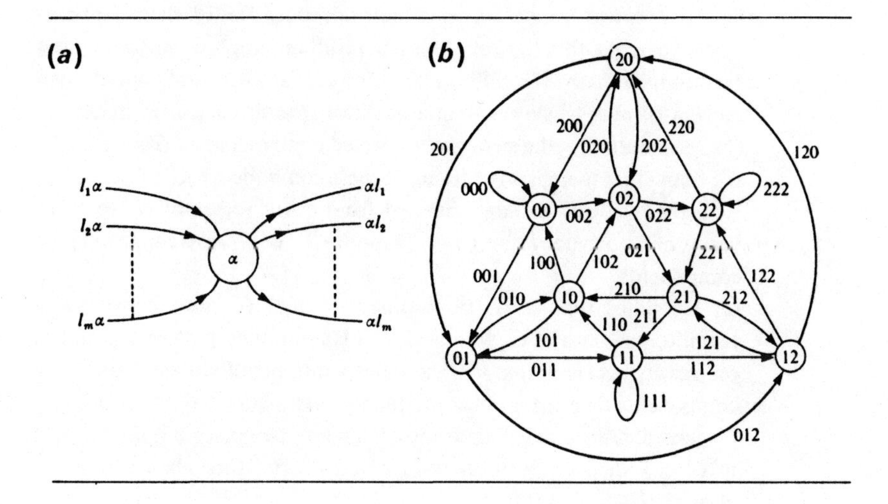

(*a*) Show that for all n and m the graph constructed according to the above rules is Eulerian. Moreover, show that an Eulerian circuit can be used to construct a circular sequence of length m^n in which no subsequence of length n is repeated by taking in turn the first letter of each edge label on the circuit. (Such a sequence for (*b*) in the above diagram is 201200011122202212110100210.)

(*b*) A metal disc, mounted on an axle, has its circumference divided into 256 equal segments. An electric current supplied via the axle will conduct radially through the disc to a contact which touches one of the circumferential segments. Some of the segments are, however, insulated. If a contact is fixed and the disc rotates then in some positions the contact will detect a current whilst in others it will not. Show that a set of eight contacts set adjacently along the circumference of the disc can just provide sufficient information to determine the orientation of the disc. This of course presupposes a suitable ordering of insulated and conducting segments. This problem has been of practical interest in telecommunications.

(*c*) In terms of m and n, how quickly can you find a de Bruijn sequence?

6.12. A major computer complex has a number of operational modes for different production work. The cost in machine down time in converting from one mode of operation i to another j is denoted by $T(i, j)$. In general we note that $T(i,j) \neq T(j, i)$. In planning a week's work schedule the operations manager notes that his installation needs to operate in each of N modes once only. He also notes that in order to minimise the total machine down time he needs to find a quickest Hamiltonian path

in a digraph G whose vertices are the operational modes and whose edges have execution times $T(i,j)$. In order to find an approximation to such a path he adopts the following sequence of heuristics:
(*a*) He removes from G, for each pair of modes i and j, the most time-costly edge of either (i,j) or (j,i). The resultant graph G' still contains a Hamiltonian path. Why? Show by example that a quickest Hamiltonian path in G' may be slower than a quickest Hamiltonian path in G.
(*b*) He determines the strongly connected components of G': $C_1, C_2, \ldots, C_j$. Show that there is an ordering of these components: $C_{i_1}, C_{i_2}, \ldots, C_{i_j}$, such that there is an edge directed from every vertex of C_{i_s} to every vertex of C_{i_t} provided that $s > t$. These are the only edges connecting the components.
(*c*) He notes that each Hamiltonian path of G' must consist of a Hamiltonian path of C_{i_1} followed by a Hamiltonian path of C_{i_2}, and so on. Justify his claim that if G' contains a number of strongly connected components then the number of Hamiltonian paths in G' will in general be considerably smaller than the number in G. Hence a quickest path in G' can then be determined in much shorter time than a quickest path in G.

In terms of both complexity and approximation, would you generally recommend employing these heuristics?

7

Colouring graphs

Our concern in this chapter is to partition or colour the vertices, edges or faces of a graph in a way dependent upon their various adjacencies. Many problems motivate these considerations, not the least of which concern scheduling and timetabling. Historically, the four-colour problem of planar maps also led to many inquiries in this area. We take a look at this problem at the end of the chapter. As we shall see, questions of partitioning and colouring are frequently intractable.

7.1 Dominating sets, independence and cliques

Board games provide ready illustrations of domination and of independence. For example an 8×8 chessboard can be represented by a graph with 64 vertices. An edge (u, v) might imply that similar chess pieces placed at the squares corresponding to u and to v would challenge one another. Any vertex adjacent to the vertex v is said to be *dominated* by v whilst any other vertex is *independent* of v.

For any graph a subset of its vertices is an *independent set* if no two vertices in the subset are adjacent. An independent set is *maximal* if any vertex not in the set is dominated by at least one vertex in it. The *independence number*, $I(G)$, of a graph G is the cardinality of the largest independent set.

A subset of the vertices of a graph is a *dominating set* if every vertex not in the subset is adjacent to at least one vertex in the subset. A *minimal* dominating set contains no proper subset that is also a dominating set. The *domination number*, $D(G)$, of a graph G is the cardinality of the smallest dominating set.

Consider again the graphical representation of a chessboard. The problem of placing eight queens on the board so that no queen challenges another, is precisely the problem of finding a maximal independent set for the graph which contains the edges (u, v) where u and v are vertices corresponding to squares in the same row or the same column or the same diagonal. There are, in fact, 92 such maximal independent sets (one is shown in figure 7.1(*a*)) and, of course, $I(G) = 8$. Another problem asks what is the minimum number of queens that can be placed on a standard chessboard such that each square is dominated by at least one queen. This problem is equivalent to finding $D(G)$ for the graph of the first problem. Figure 7.1(*b*) shows a minimal dominating set of smallest cardinality and so $D(G) = 5$.

Fig. 7.1

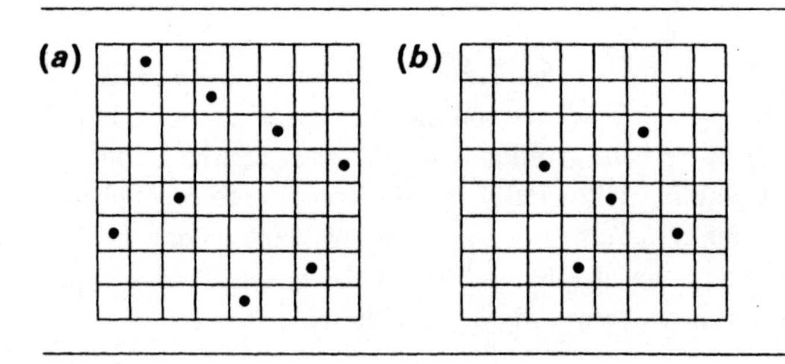

The following elementary theorem provides a relationship between $I(G)$ and $D(G)$:

Theorem 7.1. An independent set is also a dominating set if and only if it is maximal. Thus $I(G) \geqslant D(G)$.

Proof. This follows directly from the definitions. Any vertex that is not in a maximal independent set is dominated by at least one vertex in the set, hence a maximal independent set is also a dominating set. Conversely, an independent set that is also a dominating set has to be maximal because any vertex not in the set is dominated by at least one vertex in the set. ■

In connection with this theorem consider the following problem. A community wishes to establish the smallest committee to decide an issue which is of concern to a number of interested minority groups. Any individual may belong to more than one interest group and every group has to be represented. The community can be represented by a graph in which the vertices are individuals and each edge connects two individuals in the same interest group. What is required is an independent set (no interest group should be represented more than once in a smallest committee) which is also a dominating set (each group must be represented). The above theorem shows that such a choice is always possible.

We cannot present *efficient* algorithms to find $I(G)$ or to find $D(G)$ for an arbitrary graph G. In fact, for a positive integer K, the following questions are *NP*-complete:

(*a*) does G contain an independent set of size greater than K?

and

(*b*) does G contain a dominating set of size less than K?

This claim is specifically justified for (*a*) in chapter 8. Justification for (*b*), like that for (*a*), is easily obtained by showing that (*b*) is transformable from the problem of *vertex cover* which is described in chapter 8. We now describe algorithms to find $D(G)$ and to find $I(G)$.

In order that a vertex v_i is dominated we must include either v_i in a dominating set or any of the vertices $v_i^1, v_i^2, \ldots, v_i^{d(v_i)}$ which are adjacent to v_i. We can therefore (treating addition as logical *or* and multiplication as logical *and*) seek a *minimal sum of products* for the boolean expression:

$$A = \prod_{i=1}^{n} (v_i + v_i^1 + v_i^2 + \ldots + v_i^{d(v_i)})$$

in order to find the minimal dominating sets. Here, of course, if any vertex has the value *true* then it is included in the dominating set, whilst if it has the value *false* then it is excluded. For example, in connection with the graph of figure 7.2 we have that:

$$A = (a+b+d+e)(a+b+c+d)(b+c+d)(a+b+c+d+e) \times (a+d+e+f)(e+f)$$

Fig. 7.2

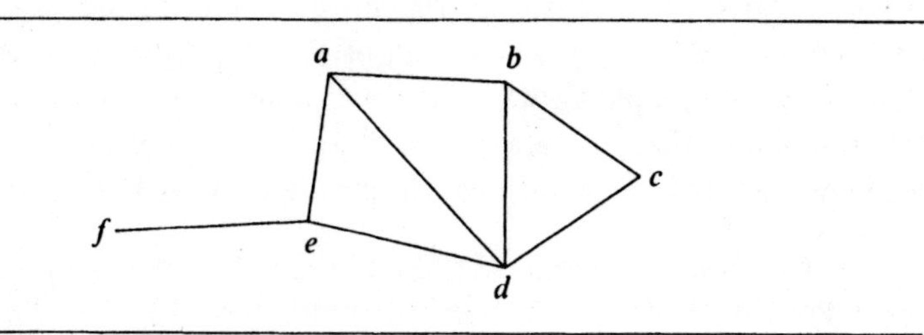

and using the identity $(u+v)v = v$ we obtain:

$$A = (a+b+d+e)(b+c+d)(e+f)$$
$$= be+de+ec+fb+fd+fac+ace$$

The seven terms in this expression respectively represent the minimal dominating sets $\{b, e\}$, $\{d, e\}$, $\{e, c\}$, ..., $\{a, c, e\}$. Five of these have the minimum cardinality of 2, so that in this case $D(G) = 2$. In general, the expression $(v_i + v_i^1 + v_i^2 + \ldots + v_i^{d(v_i)})$ contains at least two terms, so that the number of 'multiplications' involved in the evaluation of A exceeds 2^n.

Rather than directly enumerating the maximal independent sets of a graph, it is easier to enumerate the complement sets. In other words, for each maximal independent set I of the graph $G = (V, E)$, we more easily find its complement $\bar{I} = V - I$. For every edge (u, v) of the graph, $\bar{I}$ must contain u or v or both. In order to find $I(G)$ we must find the smallest sets $\bar{I}$ satisfying this condition for each edge. If we obtain a *minimum sum of products* for B, where:

$$B = \prod_{(u, v) \in E} (u + v)$$

then each term will represent a minimal set $\bar{I}$ which is guaranteed to contain at least one end-point from each edge $(u, v) \in E$. For example, for the graph of figure 7.2, we have that:

$$\begin{aligned} B &= (a+b)(a+d)(a+e)(b+c)(b+d)(c+d)(d+e)(e+f) \\ &= abce + abdf + aced + acdf + bed \end{aligned}$$

So that the graph has the maximal dominating sets:

$$\begin{aligned} &V - \{a, b, c, e\} = \{d, f\} \\ &V - \{a, b, d, f\} = \{c, e\} \\ &V - \{a, c, e, d\} = \{b, f\} \\ &V - \{a, c, d, f\} = \{b, e\} \\ &V - \{b, e, d\} = \{a, c, f\} \end{aligned}$$

The last set has the largest cardinality so that in this case $D(G) = 3$. In general, notice that evaluation of B requires $2^{|E|}$ 'multiplications'.

We consider now the rôle of *cliques* in relation to independence and dominance. A *clique* is any subgraph of $G = (V, E)$ which is isomorphic to the complete graph K_i, where $1 \leqslant i \leqslant n$ and $n = |V|$. We can always partition the vertices of a graph into cliques. Let $C(G)$ denote the number of cliques in a partition which has the smallest possible number of cliques.

Theorem 7.2. For any graph G, $I(G) \leqslant C(G)$. Also if I is an independent set and P is a partition into cliques such that $|I| = |P|$ then $I = I(G)$ and $|P| = C(G)$.

Proof. By definition, no independent set I can have more than one vertex in any clique of a partition P, hence:

$$|I| \leqslant |P|$$

Therefore:

$$I(G) = \max |I| \leqslant \min |P| = C(G)$$

If $|I| = |P|$ then the second result follows. ∎

The presence or absence of large cliques is clearly significant to the values of $D(G)$ and $I(G)$ because all the vertices in a clique are dominated by any one of its vertices. Intuitively there is a limit to the number of edges that a graph may have in order that no subgraph be a clique of specified size. We shall present a well-known theorem due to Túran[1] which provides such an upper bound. The next theorem, from which we derive Túran's, is a result due to Erdös.[2] We first need, however, to define the term *degree-majorised*. A graph G_1 is *degree-majorised* by another graph G_2 if there is a one-to-one correspondence between the vertices in G_1 and G_2 such that the degree of a vertex in G_2 is greater than or equal to the degree of its corresponding vertex in G_1. Also the *degree sequence* of a graph is defined to be the degrees of its vertices arranged in non-decreasing order.

Theorem 7.3. If G is a simple graph not containing a clique of size $(i+1)$, then G is degree-majorised by some complete i-partite graph P. Moreover, if G has the same degree sequence as P then G is isomorphic to P.

Proof. By induction on i. If $i = 1$ then G contains no edges and is degree-majorised by the 1-partite graph isomorphic to it. We thus have a basis for our induction and now assume that the theorem is true for all $i < j$. Let G be a simple graph which contains no complete subgraph K_{j+1}. We denote by G_1 a subgraph whose vertices are adjacent to a vertex u of maximum degree in G. Since G contains no K_{j+1} then G_1 contains no K_j and, therefore, by the induction hypothesis, G_1 is degree-majorised by some complete $(j-1)$-partite graph P_1. We denote by V_1 the set of vertices in G_1 and by V_2 the set of vertices $(V-V_1)$, where V is the vertex-set of G. G_2 will denote the graph with no edges but with the vertex-set V_2. Consider the *join* of G_1 and G_2 (that is, the graph obtained by drawing an edge from each vertex of G_1 to each vertex of G_2) which we denote by $J(G_1, G_2)$. In $J(G_1, G_2)$ the vertices V_2 have degree equal to the maximum degree of any vertex in G, while the vertices V_1 have at least the same degree that they have in G. Thus G is degree-majorised by $J(G_1, G_2)$. Since G_1 is degree-majorised by some complete $(j-1)$-partite graph P_1, then G is also degree-majorised by the complete j-partite graph $P = J(P_1, G_2)$. This completes the proof of the first part of the theorem.

Suppose that G has the same degree sequence as P. Then G_1 has the same degree sequence as P_1. By the induction hypothesis G_1 is then isomorphic to P_1. Also G must then have the same degree sequence as $J(G_1, G_2)$. It follows that in G each vertex in V_1 must be joined to every vertex in V_2. Thus G is isomorphic to P. ∎

Figure 7.3 illustrates the proof of theorem 7.3 for the graph G shown there. Before presenting Túran's theorem we define $T_{j,n}$ to be the complete

j-partite graph with n-vertices in which the parts are as equal in size as possible. For example in figure 7.3, H is $T_{4,9}$. In the following, $E(G)$ denotes the edge-set of G.

Fig. 7.3

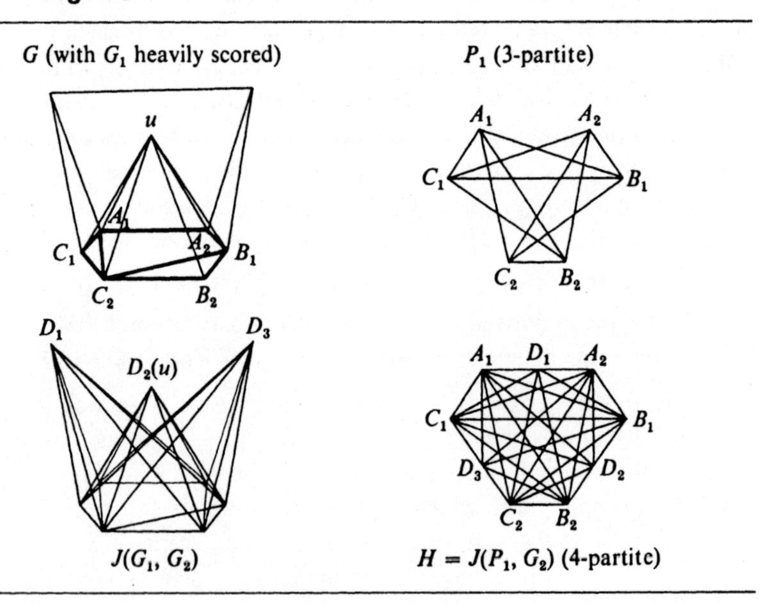

Theorem 7.4 (Túran). If G is a simple graph which does contain K_{j+1} then $|E(G)| \leqslant |E(T_{j,n})|$. Also, $|E(G)| = |E(T_{j,n})|$ only if G is isomorphic to $T_{j,n}$.

Proof. If G is a simple graph not containing K_{j+1}, then from theorem 7.3, G is degree-majorised by some complete j-partite graph, P. Hence:

$$|E(G)| \leqslant |E(P)|$$

Also, as is easily verified (exercise 7.4):

$$|E(P)| \leqslant |E(T_{j,n})|$$

and so the first part of the theorem follows. If now $|E(G)| = |E(T_{j,n})|$ then $|E(G)| = |E(P)|$ and from theorem 7.3 G and P are isomorphic. Also, since $|E(P) = |E(T_{j,n})|$ it follows (exercise 7.4) that P and $T_{j,n}$ are isomorphic. Hence G is isomorphic to $T_{j,n}$. ∎

The question of determining whether or not an arbitrary graph contains a clique greater than a given size is NP-complete. This is not surprising because, as we make use of in theorem 8.5, an independent set of a graph $G=(V,E)$ gives rise to a clique in its complement (that is, in the graph $(K_{|V|}-E)$).

7.2 Colouring graphs

In this section we first investigate the problem of colouring the edges of a graph, G, such that no two adjacent edges are similarly coloured. Such a distribution of colours is called a (*proper*) *edge-colouring* of G. Subsequently we shall be interested in colouring the vertices of G such that no two adjacent vertices are similarly coloured. In this case we refer to a (*proper*) *vertex-colouring* of G. A graph is said to be k-*edge* (or *vertex*-) colourable if a proper colouring using k colours exists.

The *edge-chromatic index*, $\psi_e(G)$, is the minimum number of colours required for a proper edge-colouring of G. Similarly, we define the *vertex-chromatic index*, $\psi_v(G)$, to be the minimum number of colours required for a proper vertex-colouring of G.

The problem of colouring the faces of (specifically planar) graphs is deferred until section 7.3. We conclude this section with a brief look at *chromatic polynomials* which are concerned with the number of ways in which a graph may be vertex-coloured.

We shall from time to time in this section and in the next refer to the *Kempe-chain argument*. Kempe published the first, but ill-fated, proof of the four-colour conjecture for plane maps in 1879. The argument concerns the recolouring of some vertices (the argument can in fact also be applied to edge-recolourings) of a proper colouring so that a different, but nevertheless proper, colouring of the graph is produced. Consider a vertex v which is coloured A. This vertex plus all the others coloured A or B which are reachable from v by paths in the graph passing only through vertices coloured A or B, constitute a component of the subgraph of G which is induced by those vertices which are coloured A or B. We denote such a subgraph by $H(A, B)$ and where appropriate we specify a component of it which includes v by $H_v(A, B)$. The Kempe-chain argument now proceeds as follows. In a proper colouring of G, those vertices in $H_v(A, B)$ coloured A can be recoloured B, and those coloured B can be recoloured A, and the result is still a proper colouring. If vertex v' is a vertex in $H(A, B)$ but not in $H_v(A, B)$, then v can be recoloured in this way without affecting the colour of v'.

7.2.1. Edge-colouring

An obvious lower bound for $\psi_e(G)$ is the maximum degree Δ, of any vertex in G. This is, of course, because the edges meeting at any vertex must be differently coloured. In fact, we shall see in due course that for any simple graph G, the following holds:

$$\Delta \leqslant \psi_e(G) \leqslant \Delta+1$$

This result is Vizing's theorem[3] which we prove after presenting theorems which are specifically concerned with bipartite and with complete graphs.

Theorem 7.5. If G is a bipartite graph then $\psi_e(G) = \Delta$.

Proof. By induction on the number of edges $|E|$. The theorem is trivially true for $|E| = 1$. We shall show that if every edge but one has been coloured with at most Δ colours, then there exists a proper colouring of G using Δ colours.

Let (u, v) be the uncoloured edge. Since there are Δ colours available, it follows that at least one colour is absent from u and that at least one colour is absent from v. If the same colour is missing at both vertices, then (u, v) can be coloured with it. Otherwise let C_1 be missing at u and let C_2 be missing at v. Of course, C_1 is present at v and C_2 is present at u. We denote by $H_u(C_1, C_2)$ the component of the two-coloured subgraph containing u. Now u and v belong to different parts of the bipartition so that *any* path from u to v within $H_u(C_1, C_2)$ must have a final edge coloured C_2. However, C_2 is missing at v and so v cannot be in $H_u(C_1, C_2)$. We can therefore interchange the colours of the edges in $H_u(C_1, C_2)$ by a Kempe-chain argument so that C_2 is absent from u as well as from v. Thus (u, v) can now be coloured C_2. ∎

Theorem 7.6. If G is a complete graph with n vertices, then

$$\begin{aligned}\psi_e(G) &= \Delta \text{ if } n \text{ is even}\\ &= \Delta+1 \text{ if } n \text{ is odd}\end{aligned}$$

Fig. 7.4

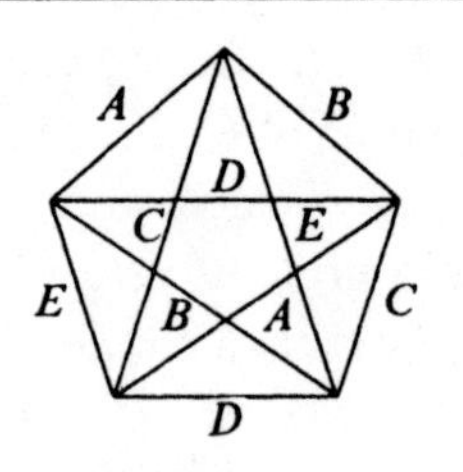

Proof. If n is odd we arrange the vertices of G in the form of a regular polygon. Figure 7.4 shows the case for $n = 5$. We colour the edges around the perimeter of the polygon using a new colour for each edge. The remaining edges are then coloured each with the same colour as the edge it is parallel with on the perimeter. Since no two edges are parallel at any vertex this must result in a proper colouring using $(\Delta+1) = n$ colours.

If G had a Δ-colouring then, since G has $\frac{1}{2}n(n-1)$ edges, there would be at least $\frac{1}{2}n$ edges with the same colour. But in a proper colouring the maximum number of edges with the same colour cannot exceed the size of a maximum-cardinality matching which is $\frac{1}{2}(n-1)$. Hence G is not Δ-colourable.

If n is even, then G can be viewed as a complete graph G' with an odd number of vertices plus an additional vertex connected to all the vertices of G'. If G' is coloured according to the process described for n odd above, then one colour is missing from each vertex. These colours are all different, so that the remaining edges of G can be coloured with the missing colours. Thus G, even n, can be properly coloured with the same number of colours as the complete graph with $(n-1)$ vertices. ■

For applications of the previous two theorems see exercises 7.5 and 7.6. These theorems show that for specific graphs $\psi_e(G)$ is equal to Δ or to $(\Delta+1)$. Vizing's theorem generalises this result.

Theorem 7.7 (Vizing). For any simple graph G:

$$\Delta \leqslant \psi_e(G) \leqslant \Delta+1$$

Proof. Since $\psi_e(G) \geqslant \Delta$ we need only show that $\psi_e(G) \leqslant \Delta+1$. We prove this by induction on the number of edges. For one edge the theorem is trivially true. We therefore suppose that all the edges of G have been properly coloured using at most $(\Delta+1)$ colours except for the edge (v_0, v_1). Since $(\Delta+1)$ colours are available there will be at least one colour missing at v_0 and at least one colour missing at v_1. If the same colour is missing at both v_0 and v_1 then this can be assigned to (v_0, v_1). We therefore assume that C_0 is missing from v_0 (but is present at v_1) and that C_1 is missing from v_1 (but is present at v_0).

We proceed to construct a sequence of edges $(v_0, v_1), (v_0, v_2), (v_0, v_3), \ldots$ and a sequence of colours $C_1, C_2, C_3, \ldots$ such that C_i is missing at v_i and such that (v_0, v_{i+1}) is coloured C_i. Let the sequences at some stage of the construction be $(v_0, v_1), (v_0, v_2), \ldots, (v_0, v_i)$ and $C_1, C_2, \ldots, C_i$. Notice that there is at most one edge (v_0, v) of colour C_i. If such a v exists and if $v \notin \{v_1, v_2, \ldots, v_i\}$ then we make v_{i+1} be v and let C_{i+1} be a colour missing at v_{i+1}, otherwise the sequence stops. Any sequence must stop with at most Δ elements. Suppose that on termination the sequences are (v_0, v_1), $(v_0, v_2), \ldots, (v_0, v_j)$ and $C_1, C_2, \ldots, C_j$. There are just two reasons why the sequences will have terminated:

(i) There is no edge (v_0, v) coloured C_j. We can then obtain a proper colouring as follows. Recolour each edge (v_0, v_i) for $i < j$ with the colour C_i. Now every edge is coloured except (v_0, v_j). But C_j is missing at both v_0 and v_j and thus (v_0, v_j) can be coloured C_j.

(ii) There exists some $k < j$ for which the edge (v_0, v_k) is coloured C_j. We obtain a proper colouring as follows. First colour each edge (v_0, v_i) for $i < k$ with the colour C_i, leave (v_0, v_k) uncoloured so that C_j is absent from v_k. Each component of the Kempe subgraph $H(C_0, C_j)$ is either a path or a circuit because at any vertex there is at most one edge coloured C_0 and at most one edge coloured C_j. Now at least one of C_0 and C_j are missing at each of the vertices v_0, v_k and v_j and so all three vertices cannot belong to the same component of $H(C_0, C_j)$. One of two circumstances must therefore occur:

(*a*) Vertex v_0 is not in the component $H_{v_k}(C_0, C_j)$. We then obtain a proper colouring as follows. Interchange the colours C_0 and C_j in $H_{v_k}(C_0, C_j)$ so that C_0 is now missing at v_k. Since C_0 is missing at v_0 we can colour (v_0, v_k) with C_0.

(*b*) Vertex v_0 is not in the component $H_{v_j}(C_0, C_j)$. We then obtain a proper colouring as follows. Recolour each edge (v_0, v_i) for $k \leqslant i < j$ with the colour C_i and leave (v_0, v_j) uncoloured. Notice that neither C_0 nor C_j is involved in this recolouring, and therefore $H(C_0, C_j)$ remains unaltered. Interchange the colours C_0 and C_j in $H_{v_j}(C_0, C_j)$ making C_0 absent from v_j. But C_0 is absent from v_0 so that (v_0, v_j) can be coloured C_0. ■

The proof of Vizing's theorem essentially embodies (see exercise 7.7) a polynomial time algorithm to obtain a proper edge-colouring of a graph using at most $(\Delta + 1)$ colours. As we shall see in chapter 8, the question of whether or not $\psi_e(G) = \Delta$, for an arbitrary graph G, is *NP*-complete. Thus the algorithm embodied in the proof might be thought of as an approximation algorithm which derives proper edge-colourings using a minimum, or very nearly a minimum, number of colours. Theorem 7.7 applies specifically to simple graphs. A more general result (which we shall not prove) also due to Vizing applies to graphs without self-loops. If M is the maximum number of edges joining any two vertices (M is called the *multiplicity* of the graph) of a graph, then:

$$\Delta \leqslant \psi_e(G) \leqslant \Delta + M$$

In fact, for any M, there exists a multi-graph such that $\psi_e(G) = \Delta + M$.

7.2.2. Vertex-colouring

Vizing's theorem (section 7.2.1) provides tight bounds on $\psi_e(G)$ for an arbitrary simple graph G. Unfortunately, as far as $\psi_v(G)$ is concerned, no theorem exists which gives such tight bounds based on simple criteria. Like $\psi_e(G)$, there is no known polynomial time algorithm to determine $\psi_v(G)$; in fact, as we prove in chapter 8, the question of whether or not a graph contains a proper vertex-colouring using less than k (a positive integer) colours is *NP*-complete. It is a simple matter to construct an exponential-time algorithm to find $\psi_v(G)$, although it requires rather more than a casual approach to get the complexity down to

$$O(|E|n(1+\sqrt[3]{3})^n)$$

as Lawler[4] has described. Throughout this section we can assume that G is a simple graph because any multi-graph has the same $\psi_v(G)$ as its underlying simple graph. The following theorem provides an obvious bound on $\psi_v(G)$.

Theorem 7.8. Any graph G is $(\Delta+1)$-vertex-colourable.

Proof. By induction on n, the number of vertices. For $n = 1$ the theorem is trivially true. If we add a vertex to the graph then this additional vertex will be attached to at most Δ other vertices and so can be coloured with the one or more colours not used by its neighbours. ■

The bound provided by theorem 7.8 can be far greater than the actual value of $\psi_v(G)$. For example, if G is planar then (see section 7.3) $\psi_v(G) \leqslant 4$ whereas G may have a vertex of arbitrarily large degree. The following theorem, due to Brooks,[5] provides only a marginally improved bound.

Theorem 7.9. If G is not a complete graph, is connected and has $\Delta \geqslant 3$, then G is Δ-vertex-colourable.

Proof. By induction on the number of vertices. Notice that if any vertex of G has degree less than Δ, then we could colour G with Δ colours by imitating the proof of theorem 7.8. Without loss of generality we can then presume that G is regular, each vertex having degree Δ. Let G have n vertices. We remove a vertex v from G so that the remaining graph has $(n-1)$ vertices and by the induction hypothesis is Δ-vertex-colourable. We suppose that all the neighbours of v are differently coloured otherwise v could be coloured with a colour missing from its neighbours. Let us denote the neighbours of v by $v_1, v_2, \ldots, v_\Delta$ and their colours, respectively, by $C_1, C_2, \ldots, C_\Delta$.

We assume that any two neighbours of v, v_i and v_j, belong to the same component of the two-coloured subgraph of G, $H(C_i, C_j)$. Otherwise v_i could be coloured C_j without affecting the colours of the other neighbours of v by a Kempe-chain argument, so freeing the colour C_i for v. We now show that every vertex of $H_{v_i}(C_i, C_j)$, apart from v_i and v_j, must be of degree 2. Starting at v_i in $H_{v_i}(C_i, C_j)$ we follow a path, not leaving any vertex by an edge along which it was approached. Suppose that we reach a vertex u (of degree Δ) which has degree greater than 2 in $H_{v_i}(C_i, C_j)$. Then there must be at least one colour absent from the neighbours of u

in G which is neither C_i nor C_j. We can recolour u with such a missing colour and so cause v_i and v_j to be in separate components of $H(C_i, C_j)$. Thus v_i could be coloured C_j and v could then be coloured C_i.

Notice that two paths such as $H_{v_j}(C_i, C_j)$ and $H_{v_j}(C_k, C_j)$ can be presumed to intersect at v_j only. Any other point of intersection, say u, would have four of its neighbours utilising only two colours. Then u would have at least one colour absent from its neighbours apart from C_j. Thus u could be coloured with a colour which is not C_i, C_j or C_k so breaking the path $H_{v_i}(C_i, C_j)$ from v_i to v_j.

We now choose any two neighbours of v, v_i and v_j, which (if such a choice is possible) are not adjacent. Let u be the vertex adjacent to v_i and coloured C_j. We can interchange the colours in $H_{v_i}(C_i, C_k), j \neq k$, without affecting the colouring of the rest of the graph. However, this leads to a contradiction because then u would be an intersection of the paths $H_{v_i}(C_i, C_j)$ and $H_{v_j}(C_k, C_j)$. Therefore we cannot choose two non-adjacent neighbours of v. Thus v and its neighbours must be $K_{\Delta+1}$ (and since G is connected this must imply that $G = K_{\Delta+1}$). This case is specifically excluded and now all possible cases have been dealt with. ■

Given that it is unlikely that the problem of finding $\psi_v(G)$ has a polynomial time solution, it is natural to think in terms of approximation algorithms. However, this problem, like a number of others (see section 7.4), seems to be unapproximable. Consider, for example, the algorithm outlined in figure 7.5. This uses the obvious heuristic of colouring the vertices in turn using the colours represented by the positive integers and such that

Fig. 7.5

```
1.  for i = 1 to n do
        begin
2.      while N_{v_i}[j] do j ← j+1
3.      for all v ∈ A(v_i) do N_v[j] ← true
4.      C(v_i) ← j
        end
```

a vertex is coloured by the integer of lowest value not used by its coloured neighbours. Such a scheme is called *sequential colouring*. The colour of vertex v_i is $C(v_i)$ and the boolean array element $N_{v_i}[j]$ is **true** if a neighbour of v_i is coloured j. Initially each $N_{v_i}[j]$ is **false**. As usual $A(v_i)$ is the adjacency list of v_i, It is easy to see that the algorithm, including any necessary initialisation, has a complexity of $O(n^2)$. The behaviour of this algorithm is highly sensitive to the order in which the vertices are coloured. For example, consider the bipartite graph $G = (V, E)$ where V is partitioned into the subsets $V_u = \{u_1, u_2, \ldots, u_k\}$ and $V_v = \{v_1, v_2, \ldots, v_k\}$, and where $E = \{(u_i, v_j) | i \neq j\}$. It is easy to see that if the vertices are coloured in the order $u_1, u_2, \ldots, u_k, v_1, v_2, \ldots, v_k$ then the graph is coloured using a minimum number of colours. But if the vertices are coloured in the order $u_1, v_1, u_2, v_2, \ldots, u_k, v_k$ then the algorithm uses $k = \frac{1}{2}n$ colours. Thus if, as usual with approximation algorithms, we define a performance ratio of C/C_0 where C is the number of colours used by the algorithm and C_0 is the optimal number, then we see that this ratio can be arbitrarily large. It is possible to modify the algorithm (see exercise 7.8) to produce an enhanced performance for many graphs. However, there are no known polynomial time algorithms for which the performance ratio is bound by a constant. The best-known performance ratio, due to Johnson[6] is $O(n/\log n)$. In fact, there is little prospect of finding a polynomial time algorithm with a good performance ratio because Garey & Johnson[7] have shown that if an approximation algorithm existed with a performance ratio of two or less, then it would be possible to find an optimal colouring in polynomial time.

We end this section by indicating a practical application of vertex-colouring. The example is a classic one concerned with timetables. A large educational institution finds itself under pressure to schedule classes so that they can all fall within acceptable teaching hours. The restricting factor is that many classes cannot be scheduled at the same time because they have to be attended by the same students. How can the designers of the timetable be certain that the scheduled lectures have been compressed into the shortest possible time? One solution is to represent the lectures as the vertices of a graph in which the edges connect vertices corresponding to lectures which cannot be scheduled at the same time. The vertex-chromatic index of this graph then represents the smallest timespan within which the lectures can be scheduled.

7.2.3. Chromatic polynomials

The idea of chromatic polynomials was introduced by Birkhoff.[8] By $P_k(G)$ we denote the number of ways of properly vertex-colouring the graph G with k colours. As we shall see, $P_k(G)$ is a polynomial in k. $P_k(G)$ is therefore referred to as the *chromatic polynomial* of G. Two simple examples are provided by the graphs of figure 7.6.

For G_1 we can colour the vertex of degree 3 first in k different ways. The remaining vertices can then each be coloured in $(k-1)$ ways. It is easy to see that for any tree with n vertices, T_n, we have $P_k(T_n) = k(k-1)^{n-1}$. Colouring the vertices of G_2 in turn provides a choice of k colours for the first, $(k-1)$ for the second and $(k-2)$ for the third. In general, for any complete graph K_n, we have $P_k(k_n) = k!/(k-n)!$ Note also that the graph with n vertices and no edges, ϕ_n, has $P_k(\phi_n) = k^n$. For $k < \psi_v(G)$, $P_k(G) = 0$ and the reader may check that this condition holds for the examples used so far.

Fig. 7.6

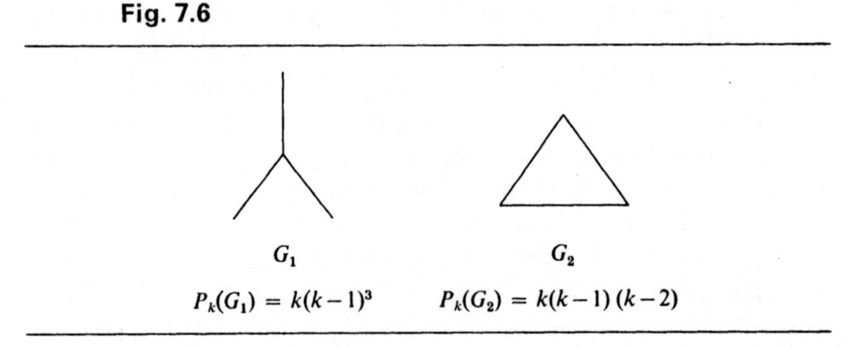

It is not necessarily an easy matter to derive $P_k(G)$ for an arbitrary graph. A useful device which provides a systematic derivation is the recursion formula of the next theorem.

Theorem 7.10. Let u and v be adjacent vertices in the graph G, then

$$P_k(G) = P_k(G-(u, v)) - P_k(G \circ (u, v))$$

where $G-(u, v)$ is derived from G by deleting the edge (u, v) and $G \circ (u, v)$ is obtained from G by contracting the edge (u, v).

Proof. Because u is adjacent to v, $P_k(G)$ consists of a count of colourings in which u is differently coloured from v. Thus all the colourings counted in $P_k(G)$ are also counted in $P_k(G-(u, v))$. However, $P_k(G-(u, v))$ includes, in addition, the number of colourings in which u and v are identically coloured, this number is specifically $P_k(G \circ (u, v))$ and so the result follows. ■

Repeated application of the recursion formula of this theorem will eventually express $P_k(G)$ as a linear combination of chromatic polynomials of graphs with no edges. We noted earlier that $P_k(\phi_n) = k^n$ and so $P_k(G)$ will be a polynomial in k. What is more, if G has n vertices, then $P_k(\phi_n)$ only appears once in the afore-mentioned linear combination. Thus $P_k(G)$ is of degree n and the coefficient of k^n is one. It is also not difficult to see that the coefficient of k^{n-1} is $(-|E|)$. This follows from the observation that removing an edge from G at each stage of a recursive evaluation of $P_k(G)$ spawns a negative term which ultimately leads to a contribution to

$P_k(G)$ from ϕ_{n-1}. We also note that the coefficient of k^0 in $P_k(G)$ must be zero, because if $k = 0$ then we must have $P_k(G) = 0$.

The formula of theorem 7.10 may also be applied in the following form:

$$P_k(G) = P_k(G+(u, v))+P_k((G+(u, v)) \circ (u, v))$$

Fig. 7.7

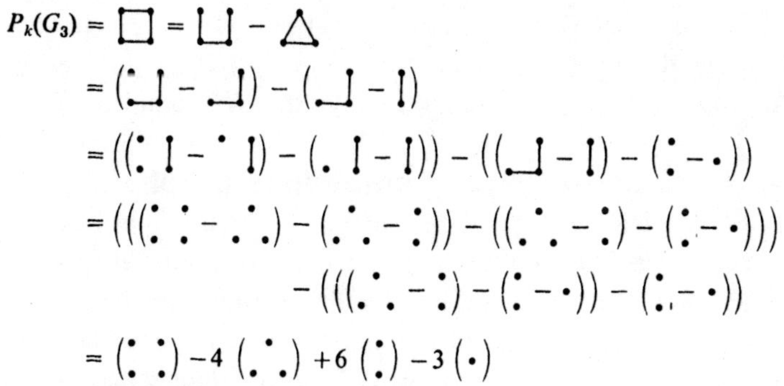

$P_k(G_3) =$

$= P_k(\phi_4)-4P_k(\phi_3)+6P_k(\phi_2)-3P_k(\phi_1)$

$= k(k-1)(k^2-3k+3)$

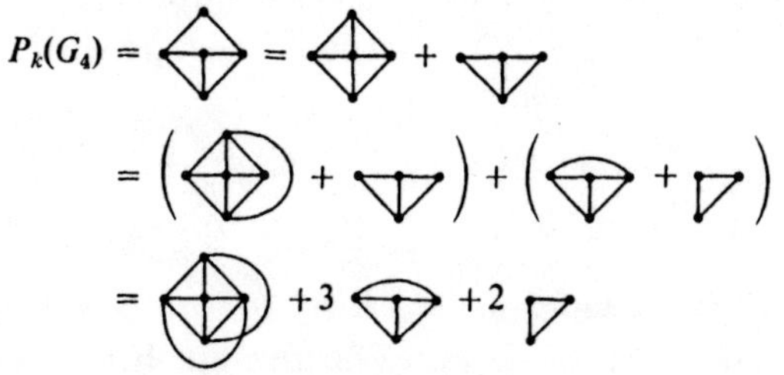

$P_k(G_4) =$

$= P_k(K_5)+3P_k(K_4)+2P_k(K_3)$

$= k(k-1)(k-2)(k^2-4k+5)$

Recursive evaluation of $P_k(G)$ using this form will eventually express $P_k(G)$ as a linear combination of chromatic polynomials of complete graphs. If G has a large number of edges then this mode of solution will evaluate $P_k(G)$ more quickly than the former method. In figure 7.7 we illustrate the two methods of evaluating $P_k(G)$. Obvious convenience is made of representing the chromatic polynomial of a graph by the graph itself. Also, whenever more than one edge arises between two vertices only one edge is retained. Obviously, $\psi_v(G)$ is the smallest value of k for which $P_k(G) > 0$. Thus $\psi_v(G_3) = 2$ and $\psi_v(G_4) = 3$.

The original motivation for studying chromatic polynomials was to seek a solution to the four-colour problem of planar maps. In terms of chromatic polynomials the four-colour conjecture (see section 7.3) would be proven true if for any planar graph (which is the *dual* (see section 3.3.1) of any planar map) G:

$$P_4(G) > 0$$

In the event such a method of solution has not been found.

It is unlikely that $P_k(G)$ can be found in polynomial time because this would imply that an efficient determination of $\psi_v(G)$ existed. This in turn would provide an efficient solution to any other NP-complete problem.

7.3 Face-colourings of embedded graphs

This section is largely concerned with planar graphs and the four-colour conjecture which was eventually proved correct by Appel & Haken. The conjecture that four colours are sufficient to colour the regions of a plane map (that is, the faces of a graph embedded in the plane) so that bordering regions (adjacent faces) are differently coloured became perhaps the best-known unsolved problem in mathematics. It became so because it withstood the onslaught of many mathematicians for over 120 years.

Although our concern is specifically with planar embeddings, we note in passing that for maps of genus $g \geqslant 1$ Heawood[10] has shown that the following numbers of colours are sufficient:

$$\left\lceil \frac{7+\sqrt{(1+48g)}}{2} \right\rceil$$

Proof of this formula does not unfortunately carry over for $g = 0$. Also, Heawood unwittingly presumed the necessity for this number of colours but proof of this was not obtained until Ringel & Youngs[11] published their work. For a discussion of this see chapter 3 of Beineke & Wilson.[12]

In section 7.3.2 we indicate the lines along which Appel & Haken eventually provided a proof of the four-colour conjecture. Because of the complexity of their work it is not possible to provide a detailed description. Fortunately, we can easily show that five colours are sufficient to provide a proper face-colouring for any planar graph. This we do in the following section.

7.3.1. The five-colour theorem

Kempe[13] published what seems to have been the first attempted proof of the four-colour conjecture. Although Kempe's work contained a flaw which Heawood[10] pointed out, it contained a valuable contribution which formed the basis of many later attempts to solve the problem including Appel & Haken's successful attempt. Kempe marshalled the following ideas:

(*a*) As we described at the end of section 3.3, showing that $\psi_v(G) \leqslant 4$ for any simple planar graph is equivalent to proving the four-colour conjecture.

(*b*) In colouring the vertices of a simple planar graph it is sufficient to consider plane triangulations only. A *plane triangulation* is obtained from an embedding of any planar graph G by adding edges so as to divide each non-triangular face into triangles. Figure 7.8 shows the addition of (dashed) edges to form a plane triangulation. Clearly, any planar embedding G is a subgraph of some plane triangulation T, so that a proper vertex-colouring of T will be a proper vertex-colouring of G.

Fig. 7.8

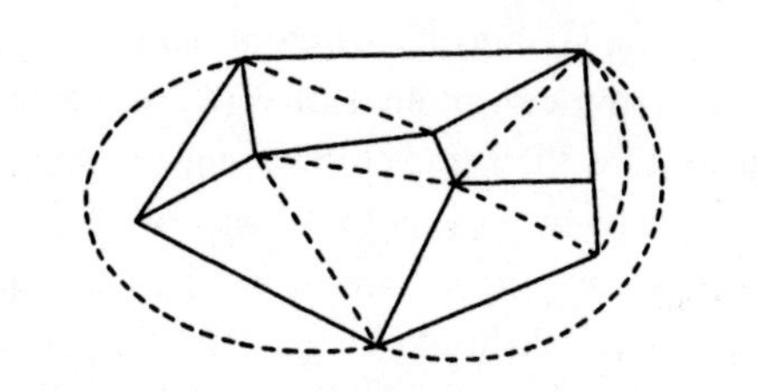

(*c*) As we proved for corollary 3.3, every planar graph contains at least one vertex of degree at most 5. Hence, any plane triangulation contains one of the *configurations* (subgraphs) illustrated in figure 7.9. Notice that a plane triangulation cannot contain a vertex of degree 1.

Fig. 7.9

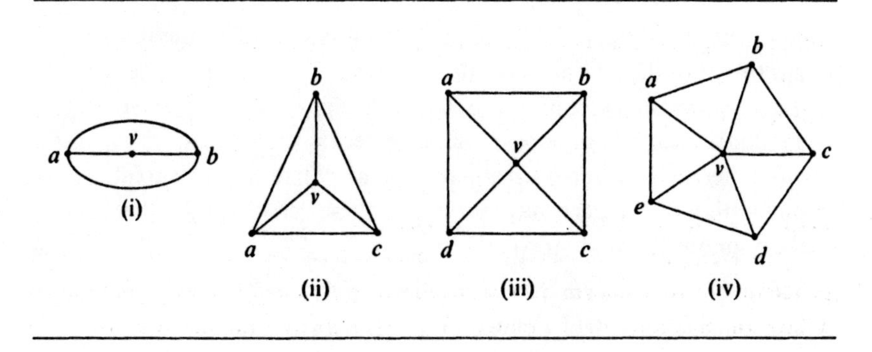

(*d*) Kempe-chain arguments. This style of reasoning was described at the beginning of section 7.2.

Kempe's attempt at a proof of the four-colour conjecture used induction on the number of vertices of a plane triangulation. For $n \leqslant 4$ the conjecture is clearly true. Suppose that $n > 4$. The plane triangulation will contain one of the configurations of figure 7.9. If v is removed from the triangulation, T, then the remaining graph is 4-vertex-colourable by the induction hypothesis. Given such a colouring then we are required to replace and colour V without the use of an additional colour. For configurations (i) and (ii) of figure 7.9 this can be done trivially by assigning to v a colour not utilised at a, b or c. We presume then that T does not contain the configurations (i) and (ii). Consequently it contains (iii) or/and (iv).

Kempe used the following argument to deal with configuration (iii). We presume that the vertices a, b, c and d are all differently coloured, otherwise v could be replaced and coloured with an unused colour. Let a, b, c and d be then coloured C_1, C_2, C_3 and C_4. It cannot be the case that vertices a and c belong to the same component of $H(C_1, C_3)$ and *at the same time* vertices b and d belong to the same component of $H(C_2, C_4)$. Clearly, such a supposition would lead to these components of $H(C_1, C_3)$ and $H(C_2, C_4)$ having at least one vertex, u, in common. The colour of u cannot be (C_1 or C_3) and (C_2 or C_4). Without loss of generality, we may assume that a is *not* in the same component of $H(C_1, C_3)$ as c and hence, by a Kempe-chain argument, vertex a may be coloured C_3 without affecting the colours of b, c and d. This makes the colour of C_1 available for v.

Unfortunately, Kempe's treatment of the configuration (iv) of figure 7.9 contained an error. His argument relied upon two simultaneous colour changes which, as Heawood showed, can cause two adjacent vertices to become similarly coloured. Nevertheless, Heawood was able to salvage the following result as the *five-colour theorem*.

Theorem 7.11. Every planar map is 5-face-colourable.

Proof. This exactly parallels Kempe's attempted proof of the four-colour conjecture. For the first three configurations of figure 7.9, v can be replaced and coloured with a colour not used at a, b, c or d. As far as configuration (iv) is concerned the only non-trivial case occurs when the neighbours of v are all differently coloured. In a manner exactly like Kempe's treatment of configuration (iii), we can recolour one of these vertices and in the process make a colour available for v. ■

The proof of theorem 7.11 embodies a polynomial time algorithm to five-colour the vertices of a planar graph. It is an easy matter (exercise 7.11) to describe an $O(n^2)$ implementation. However, it is possible (see, for example, Chiba *et al.*[14]) to describe a linear-time implementation of five-colouring.

7.3.2. The four-colour theorem

As we stated earlier, the basis of Appel & Haken's proof of the four-colour conjecture (which we can now properly call the four-colour theorem) can be traced back to Kempe's attempt. After Kempe many other researchers also contributed in the direction of this successful proof. Of course Appel & Haken's achievement was a major one and involved a great extension both qualitatively and quantitatively from this simple base.

In a text of this kind it is not appropriate, let alone possible, to elucidate every detail of a proof which in one area made massive use of computer time and which in another involved a long period of trial and error and of insight gained from the results and performances of computer programs. However, what we can do is to outline the deceptively simple concepts behind the proof and then hopefully try to give some idea of the technical difficulties involved in pursuing them.

The object of the proof was to show, like Kempe's attempt, that every plane triangulation has a 4-vertex-colouring. The two essential concepts behind the proof are those of *unavoidable sets* and *reducible configurations*. Before defining these we need to explain what is meant by a configuration. A *configuration* consists of part of a plane triangulation contained within a circuit. This circuit is called the *ring bounding the configuration* and the number of vertices in the circuit is called the *ring-size* of the configuration. Figure 7.9 shows some configurations with respective ring-sizes of 2, 3, 4 and 5.

An unavoidable set is a set of configurations such that every plane triangulation must contain at least one of the configurations in the set. Thus figure 7.9 provides one example of an unavoidable set.

A configuration is said to be reducible if it cannot be contained in a triangulation which would be a smallest counterexample to the four-colour conjecture. For example, we saw in section 7.3.1 that a counter-example would not include configurations (i), (ii), and (iii) of figure 7.9. If Kempe had been able to show that configuration (iv) of that diagram was also reducible then his proof would have been complete.

The starkest description of Appel & Haken's proof is that they were able to find an unavoidable set of reducible configurations. This description, however, belies the fact that the proof required a great deal of effort and ingenuity in order to avoid intractable computation. It would be incorrect to presume that their task was simply divided into two parts, that of finding an unavoidable set and that of proving that every configuration in the set was reducible. In fact, both parts were made to play a strongly interdependent rôle in the development of the final set of reducible configurations.

The search for alternative unavoidable sets has a long history. In 1904 Weinicke published the unavoidable set shown in figure 7.10. Others were published by Franklin in 1922 and Lebesque in 1940. Appel & Haken's

Fig. 7.10

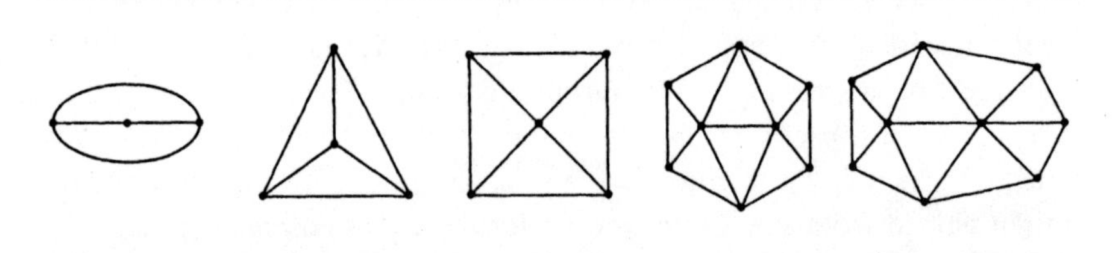

method for proving that a set of configurations is unavoidable was a development of the so-called method of discharging presented by Heesch in 1969. The principle behind this method is as follows. Each vertex is assigned a number $(6-i)$ where i is the degree of the vertex. This number is called the *charge* on the vertex. For any plane triangulation we have from corollary 3.3 that:

$$\sum_i (6-i)\, n(i) = 12$$

where $n(i)$ is the number of vertices with degree i. Hence the total charge for any plane triangulation must be 12. Given a set of configurations, S, we suppose that there exists a triangulation, T, not containing any configuration in S. If we can redistribute the charge in T (without creating or destroying charge) such that no vertex ends up with a positive charge then we have a contradiction. The total charge must be positive and so the assumption that S is not an unavoidable set is proved false. The difficulty, in general, is how to redistribute the charge. We can demonstrate the kind of technique with a rather simple example.

We shall show that Wernicke's set of configurations shown in figure 7.10 in unavoidable. T is then a plane triangulation with no vertices of degree 2, 3 or 4 and no vertex of degree 5 adjacent to a vertex of degree 6 or less. Notice that only vertices of degree 5 are positively charged (with one unit) in the first instance. We now allow each vertex of degree 5 to discharge one-fifth of a unit of charge to each of its neighbours. In this way every vertex ends up with a non-positive charge, because any vertex with degree i ($\geqslant 7$) will have at most $\sim \frac{1}{2}i$ neighbours of degree 5. Thus we have a contradiction of the type described earlier.

As a result of using a particular discharging procedure Heesch thought that the four-colour problem could be reduced to considering a finite set of configurations. In fact, he explicitly exhibited a set of $\sim$ 8900 configurations. The ring-size (up to 18) of some of these was too large for them to be tested for reducibility in a practical length of time. Heesch also observed that in investigating the reducibility of some configurations, certain features (so-called *reduction obstacles*) appeared to prevent reduction. Appel & Haken's task was to find a set of configurations with manageable ring-size (in the event 14 was the largest) and which avoided the reduction obstacles.

The study of reducibility, like that of unavoidable sets, has a long history. Birkhoff wrote an important paper in 1913 in which, amongst other things, he proved that the so-called Birkhoff diamond (figure 7.11 (*a*))

Fig. 7.11

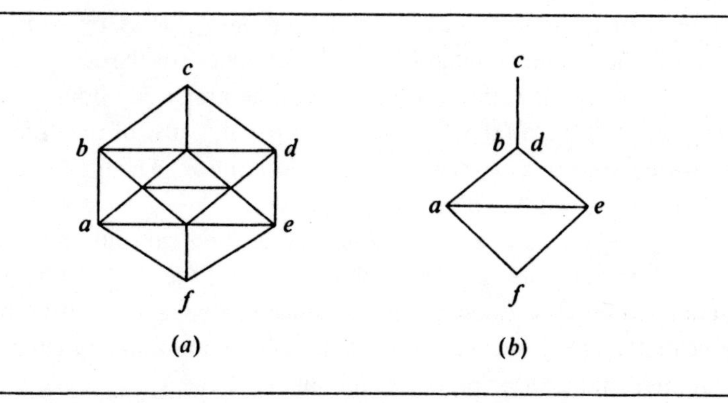

is reducible. Thousands of reducible configurations are now known following the interest of many mathematicians. The sort of method used by Heesch (which is a refinement of Birkhoff's) can be briefly described as follows. The object is to show that a four-colouring of a plane triangulation T, not containing a particular configuration, C, can be extended to include the configuration. Any colouring of T, for this purpose, can be represented by the way it distributes the available four colours amongst the vertices in the ring bounding the configuration C. A list of all possible permutations of the colours on the ring is constructed. Several of these permutations allow a colouring of the configuration immediately and can therefore be removed from the list of permutations to be considered. Next, Kempe-chain arguments can be applied to the remaining permutations and this converts some of them into permutations that have already been discarded and so they can be discarded as well. Configurations that can be entirely dealt with in this way are called *D*-reducible. If the configuration can be dealt with by considering a subset of permutations, in fact a subset corresponding to replacing C with a smaller configuration, then the original configuration is called *C*-reducible. An example for the Birkhoff diamond is shown in figure 7.11 (*b*) where, as required, the rest of the graph would be left intact. Appel & Haken's computer programs utilised these methods, abandoning particular configurations when proof of reducibility proved too lengthy, at the expense of introducing one or more further configurations.

Eventually, an unavoidable set of approximately 1500 reducible configurations was constructed. This was achieved after a long period of trial and error, involving much empirical adjustment to a complicated discharging algorithm depending on the interplay between the developing unavoidable set and the discharging procedure. Appel & Haken developed such a strong intuitive sense for what was likely to be successful that they were eventually able to enact the discharging process by hand and so construct the final unavoidable set without the use of the computer. Appel is on record as estimating that it would take about 300 hours on a large computer to check all the details of their proof, many of the difficulties of which we have skated over in this brief description. The final unavoidable set used in the proof is illustrated in [17].

Appel & Haken's proof of the four-colour theorem could not have been achieved without the computer. Unfortunately, the sheer effort and time required to check every detail mitigate against wide verification by many other mathematicians. No doubt there will be a continuing effort to seek alternative and shorter proofs of the theorem.

7.4 Summary and references

As we shall confirm in chapter 8, many graph colouring problems are *NP*-complete. The problem of vertex-colouring is particularly intransigent because, as we saw in section 7.2.2, it seems that we cannot even find a useful approximation to $\psi_v(G)$ in polynomial time. The problem of finding the independence number of a graph is similarly non-approximable (Garey & Johnson,[7] see also exercises 7.10 and 8.15).

Practical applications of the material of this chapter can be found in exercises that follow as well as in the chapter.

For further general reading see chapters 12–15 of Berge,[19] chapter 12 of Harary,[20] chapter 4 of Busacker & Saaty[21] and chapters 6–8 of Bondy & Murty.[18]

The book by Ore[9] provides a fund of background information on the four-colour problem. The articles by Appel & Haken[15,16,17] describe their solution, as does chapter 4 of Beineke & Wilson.[12]

[1] Túran, P. 'An extremal problem in graph theory', *Mat. Fiz. Lapok*, **48**, 436–52 (1941).
[2] Erdös, P. 'On the graph theorem of Túran', *Map. Lapok*, **21**, 249–51 (1970).
[3] Vizing, V. G. 'On an estimate of the chromatic class of a *p*-graph', *Diskret. Analiz.*, **3**, 25–30 (1964).
[4] Lawler, E. 'A note on the complexity of the chromatic number problem', *Rapport de Recherche*, No. 204, IRIA (December 1976).
[5] Brooks, R. L. 'On colouring the nodes of a network', *Proc. Cambridge Philos. Soc.*, **37**, 194–7 (1941).
[6] Johnson, D. S. 'Worst case behaviour of graph colouring algorithms', *Proc. 5th Southeastern Conference on Combinatorics, Graph Theory and Computing*. Utilas Mathematica Publishing, Winnipeg, pp. 513–27 (1974).
[7] Garey, M. R. & Johnson, D. S. *Computers and Intractability – A Guide to the Theory of NP-Completeness*. W. H. Freeman, San Francisco (1979).
[8] Birkhoff, G. D. 'A determinant formula for the number of ways of colouring a graph', *Ann. of Math.* **14**, 42–6 (1912).
[9] Ore, T. *The Four-Colour Problem*. Academic Press (1967).
[10] Heawood, P. J. 'Map-colour theorem', *Quart. J. Pure Appl. Maths.*, **24**, 332–8 (1890).
[11] Ringel, G. & Youngs, J. W. T. 'Solution of the Heawood map-colouring problem', *Proc. Nat. Acad. Sci., USA*, **60**, 438–45 (1968).
[12] Beineke & Wilson (eds.). *Selected Topics in Graph Theory*. Academic Press (1978).
[13] Kempe, A. B. 'On the geographical problems of the four colours', *Amer. J. Maths.*, **2**, 193–200 (1879).
[14] Chiba, N. Nishizeki, T. & Saito, N. 'A linear algorithm for five-colouring a planar graph', *Lecture notes in Computer Science*, **108**, *Graph Theory and Algorithms*. Springer-Verlag, 9–19 (1980).
[15] Appel, K. & Haken, W. 'The solution to the four-colour problem', *Scientific American*, **27** (4), 108–21 (1977).
[16] Appel, K. & Haken, W. 'Every planar map is four-colourable', part I: 'Discharging', *Illinois J. Maths*, **21**, 429–90 (1977).
[17] Appel, K. & Haken, W. 'Every planar map is four-colourable', part II: 'Reducibility', *Illinois J. Maths*, **21**, 491–567 (1977).
[18] Bondy, J. A. & Murty, U. S. R. *Graph Theory with Applications*. The MacMillan Press (1976).
[19] Berge, F. *Graphs and Hypergraphs*. North-Holland (1973).
[20] Harary, T. *Graph Theory*. Addison-Wesley (1969).
[21] Busacker, P. & Saaty, M. *Finite Graphs and Networks*. McGraw-Hill (1965).
[22] Isaacs, R. 'Infinite families of non-trivial graphs which are not Tait-colourable', *Amer. Math. Monthly*, **82**, 221–39 (1975).

EXERCISES

7.1. Show that any graph with n vertices and at least $\lceil \frac{1}{4}n^2 \rceil$ edges contains a triangle.
(Use theorem 7.4.)

7.2. In a group of eight people one person knows three others, two know four others, three know six others, while the remaining two each know two others. Show that there must be a group of three mutual acquaintances.
(Use theorem 7.4.)

7.3. A national radio service transmits to a population residing in townships distributed throughout the country. Each town is within radio transmission distance of at least one other. If radio transmitters are to be located in the towns, the problem of economically siting them is one of finding a minimal dominating set. For purposes of reliability, however, the engineers wish to have two sets of transmitters, each set being operated as a unit. If one unit breaks down, then the whole population is still to be serviced by the second unit. Show that this can be done within the constraint that no two transmitters are located in any one township and that it may not be necessary to have a transmitter in every town.
(The problem is equivalent to showing that in every connected graph there are two disjoint minimal dominating sets. If V is the vertex-set and X is a minimal dominating set, show that $(V-X)$ is a (not necessarily minimal) dominating set.)

7.4. P is a complete m-partite graph with n vertices. $T_{m,n}$ is the complete m-partite graph with n vertices in which the numbers of vertices in each part are as equal as possible. Show that:
$|E(P)| \leqslant |E(T_{m,n})|$
and that if equality holds then P and $T_{m,n}$ are isomorphic. (The number of vertices in any part of $T_{m,n}$ differs by no more than one from the number in any other part. This is not true for P if $T_{m,n}$ and P are not isomorphic. Show then that $|E(P)| < |E(T_{m,n})|$.)

7.5. A *Latin square* is an $N \times N$ matrix in which the entries are integers in the range 1–N. No entry appears more than once in any row or any column. Justify the following construction of a Latin square, T.
Form a complete bipartite graph G, each part having N vertices. Properly edge-colour G (note theorem 7.5) using a minimum number of colours. In such a colouring associate the edge-colours of $G = (V, E)$ with the column indices of T and associate the indices of the vertices in one part of G (let these be V_1 and let $V_2 = V - V_1$) with the row indices of T. Then $T(i,j)$ is assigned the index of the vertex we arrive at in V_2 by following the edge coloured j from the ith vertex in V_1.

7.6. There are $2N$ contestants in a chess tournament. No contestant plays more than one match in a day and must, in the course of the tournament, play every other player. Justify the following scheme to complete the tournament in $(2N-1)$ days.

Properly colour the edges (note theorem 7.6) of the complete graph K_{2N} using a minimum number of colours. Label the colours 1–$(2N-1)$. Let the vertices of K_{2N} represent the players. On the ith day those players connected by edges coloured i are drawn together.

7.7. Describe, in a style similar to that used for algorithmic description throughout this text, a polynomial time algorithm to properly edge-colour a graph using at most (note the proof of theorem 7.7) $\Delta+1$ colours, where Δ is the maximum degree of any vertex of the graph. Make the implementation as efficient as you can.

7.8. Modify the sequential colouring algorithm of figure 7.5 so that it interchanges two colours C_i and C_j in any two-coloured subgraph of the coloured portion of the graph if doing so will avoid the use of a new colour for the next vertex v_k to be coloured. This will be the case if each neighbour of v_k which is coloured C_i does not belong to a same component of the Kempe subgraph $H(C_i, C_j)$ as any neighbour of v_k which is coloured C_j.

(*a*) Show that the modified algorithm will produce a proper two-colouring for any bipartite graph irrespective of the order in which vertices are coloured.

(*b*) Show that the modified algorithm can utilise, for particular colouring sequences, up to k colours for graphs $G' = (V', E')$ where $V' = \{u_i, v_i, w_i | 1 \leqslant i \leqslant k\}$, $E' = \{(u_i, v_j), (u_i, w_j), (v_i, w_j) | i \neq j\}$ and $k \geqslant 3$. Notice that $\psi_v(G') = 3$ and hence, as for the unmodified algorithm, the performance ratio can be arbitrarily large.

7.9. *Prove the following lemma due to Isaacs*[22]

$G = (V, E)$ is a 3-regular graph with a proper 3-edge-colouring. V' is any subset of vertices and E' is the set of edges connecting vertices in V' to vertices in $(V-V')$. If the number of edges coloured i in E' is K_i, $i = 1, 2, 3$, then the K_i are either *all* even or are *all* odd.

(Consider the components of $H(C_i, C_j)$ all of which are circuits.)

7.10. Show (by example) that the following algorithm to find an approximation I to a maximum independent set I_0 has an arbitrarily large performance ratio $(|I_0|/|I|)$.

1. $I \leftarrow \varnothing$
2. **for** $i = 1$ **to** n **do**
3. **If** v_i is not adjacent to any $v \in I$
4. **then** $I \leftarrow I \cup \{v_i\}$

Demonstrate that the performance of the algorithm is highly sensitive to the order in which the vertices are labelled.

7.11. Use the proof of the five-colour theorem, theorem 7.11, to construct a polynomial time algorithm to properly 5-vertex-colour any planar graph. The complexity of your algorithm will probably be quadratic, however note the paper of Chiba *et al.*[14]

7.12. (*a*) Show that a planar graph G has a 2-face-colouring if and only if G is Eulerian.

(Show that the *dual* of G is bipartite and that any bipartite graph has an Eulerian dual.)

(*b*) Show that every planar Hamiltonian graph has a 4-face-colouring. (Any Hamiltonian circuit divides the plane into two regions. Consider using two colours for the faces in either region.)

7.13. N committee members sit at a round table. K different organisations have representatives on the committee. How many ways can the members be seated subject to the constraint that each organisation has a varying membership of at least one representative and when it has more than one then no two of them may sit next to each other?

(The answer is provided by the chromatic polynomial of the circuit of length n, C_n:

$P_K(C_n) = (k-1)^n+(-1)^n\,(k-1))$

7.14. Show that the following two statements are equivalent:

(*a*) Every simple planar graph is 4-face-colourable.

(*b*) Every simple 3-regular, 2-edge-connected planar graph is 3-edge-colourable.

(Show that (*a*) implies (*b*) as follows. Let G be a simple cubic, 2-edge-connected planar graph. According to (*a*) it has a 4-face-colouring, let such a colouring use the colours A, B, C and D. Assign the colours α, β and γ to the edges as follows:

α if the edge separates faces coloured (A and B) or (C and D)

β if the edge separates faces coloured (A and C) or (B and D)

γ if the edge separates faces coloured (A and D) or (B and C)

This is easily seen to be a proper 3-edge-colouring of G.

Show that (*b*) implies (*a*) as follows. An equivalent statement to (*a*) is that any plane triangulation is 4-vertex-colourable. But any simple planar graph with triangular faces is the dual of some cubic, 2-edge-connected graph. We need, therefore, only show that (*b*) implies that any cubic, 2-edge-connected graph is 4-face-colourable. Let G be such a graph with a 3-edge-colouring using colours α, β and γ. The two-coloured subgraph $H(\alpha, \beta)$ of G is 2-regular and so (see 7.12(*a*)) has a 2-face-colouring using colours a and b, say. Similarly, $H(\alpha, \gamma)$ has a 2-face-colouring using colours c and d. Each face of G lies within a face of $H(\alpha, \beta)$ and a face of $H(\alpha, \gamma)$ and can therefore be assigned a pair of indices, namely (a or b) and (b or d). In fact, each face of G is an intersection of a face of $H(\alpha, \beta)$ and a face of $H(\alpha, \gamma)$ so that any two adjacent faces of G differ in at least one index. Thus the following assignments provide a proper 4-face-colouring using colours 1, 2, 3 and 4.

$1 \equiv (a, c)$, $2 \equiv (a, d)$, $3 \equiv (b, c)$, $4 \equiv (b, d)$).

A proper 3-edge-colouring of a cubic graph is called a *Tait colouring*. In 1880 Tait gave a proof of the four-colour conjecture for planar maps on the assumption that every cubic 2-edge-connected planar graph is Hamiltonian. However, in 1946 Tutte showed that this was an invalid assumption by constructing the non-Hamiltonian cubic 3-connected graph shown below.

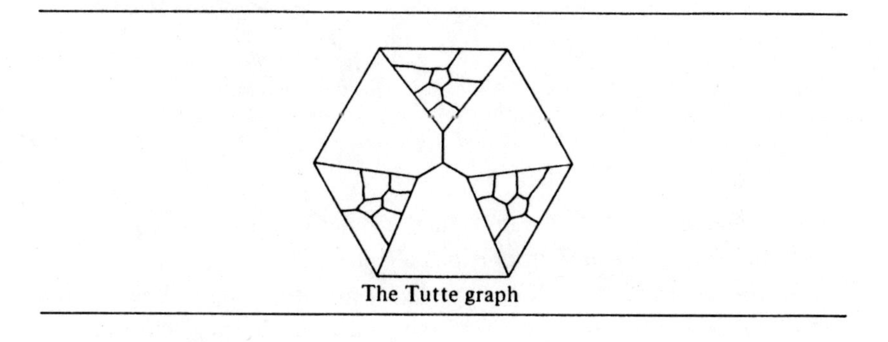

The Tutte graph

8

Graph problems and intractability

In this chapter we provide a formal framework for the concept of algorithmic efficiency used in previous chapters.

We re-emphasise the difference between those algorithms whose execution times are bounded by a polynomial in the problem size and those which are not. Furthermore we introduce a broad class of problems, the so-called *NP*-complete problems, which are widely believed to be inherently intractable. This belief is largely based on the circumstantial evidence that, despite the expenditure of much effort in the search for efficient algorithms, not one is known for any member of this class. Moreover, if such an algorithm was known for any one of these problems, then such an algorithm would exist for any one of the others.

We shall see that every problem, for which we were unable to provide an efficient solution earlier, is *NP*-complete.

8.1 Introduction to *NP*-completeness

Previously we have found it unnecessary to have a formal definition of the concept of an algorithm. Any algorithm we introduced consisted of a set of informally expressed instructions. In every case this description could easily be embodied in a computer program. We shall have a need later to be more precise, taking the definition of an algorithm to be that embodied in *Church's thesis* that there is an algorithm to solve a particular problem if and only if it can be solved by a *Turing machine* which halts for every input. Turing machines provide a model of computation which we describe shortly. The usefulness of this model lies within its simplicity which greatly facilitates theorising about algorithms. A number of other formal models are commonly used to prove complexity theorems, some of these (unlike the Turing machine) are random access machines and are therefore fairly realistic. All these models are equivalent with respect to polynomial time complexity, each able to simulate a computation on the other at (low order) polynomial time cost. See, for example, chapter 1 of Aho, Hopcroft & Ullman.[8] This allows the viewpoint that those problems which we have 'shown' to have efficient algorithms in earlier chapters retain their efficiency in terms of the Turing machine model. By the same token, we can continue to use informal arguments to demonstrate efficiency, using our previous method to describe algorithms, when it suits our purpose.

Conveniently, polynomials bring an algebraic advantage in what theory follows because as a class of functions they are closed under the operations of addition, multiplication and composition. In other words, a polynomial results from the addition or the multiplication of two polynomials and from taking a polynomial of a polynomial. This can be useful from two points of view. First, a complex algorithm can often be subdivided into a number of smaller polynomial time components. If the complexity of the original algorithm is bounded by the addition, multiplication and composition of the complexities of its components, then it will itself be of polynomial time complexity. Secondly, it is often the case that one problem may be solved by transforming it into another. If we have efficient algorithms both for the transformation process and for the second algorithm, then it follows that we have an efficient algorithm for the initial problem. This idea will be of particular interest in section 8.1.2 where we describe the notion of *NP*-completeness.

8.1.1. The classes P and NP

In order to define the classes of problems *P* and *NP* we need first to complete our formal definition of algorithm by a description of Turing machines.

A Turing machine (*TM*) carries out its computation on an infinite *tape* which is divided into cells along its length. At any one time a cell contains a single symbol or is blank. At the outset of a computation a finite set of contiguous cells contains an encoding of the input to the computation and all other cells are blank. The computation then proceeds by the repetition of a cycle of actions involving the *tape head.* This device has a finite number of internal states. The actions within a cycle are as follows: the tape head reads the contents of a single tape cell, then depending upon what it has read and which state it is in it replaces the contents of that cell with a new symbol (this may in fact be identical to the old symbol), enters a new state (which might be its old state) and moves one cell to the right or to the left before starting on the next cycle. The specific changes occurring in any one cycle are determined by a *quintuple*, or instruction, written as follows:

$$(q_i, t_i, q_f, t_f, m)$$

where q_i is the current state, q_f the new state, t_i is the current tape symbol and t_f the new tape symbol. Whether or not the tape head moves to the right or to the left depends upon whether, respectively, m has the value $+1$ or the value (-1). Since the set of states, Q, and the set of tape symbols (called the *tape alphabet*), T, are finite, any computation can be completely specified by a finite set of quintuples. Such a set constitutes a (non-sequential) program for the *TM*. For the moment we presume that a computation is deterministic in the sense that for any pair (q_i, t_i) precisely one quintuple exists to determine (q_f, t_f, m); the computation is then said to be performed by a *deterministic TM* or *DTM*. At the outset of a computation a *TM* is conventionally in its *initial state* q_0, and a specified tape cell is being scanned. If the tape cells are labelled ..., $C(-2)$, $C(-1)$, $C(0)$, $C(1)$, $C(2)$, ..., we shall presume that $C(0)$ is always this initially scanned cell. The computation halts when the *TM* enters one of a set of final states, $F \subset Q$. The result of the computation is then either encoded within the symbols that remain on the tape or is indicated by the particular halt state that the *TM* is in. The latter case is especially suitable for *decision problems* which simply require a 'yes' or 'no' answer. As an example of a *TM* we now make use of just such a problem.

Fig. 8.1

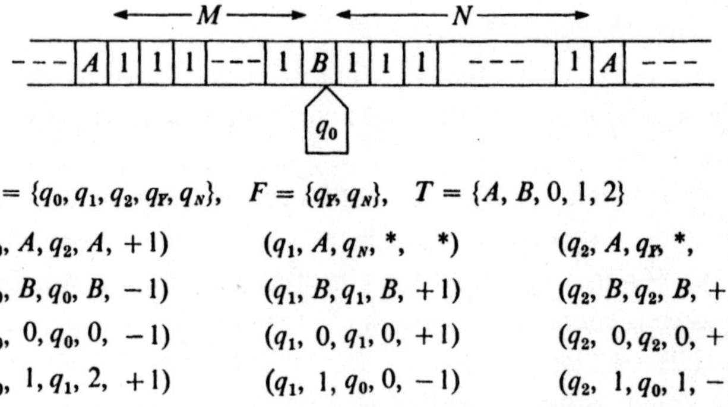

$\longleftarrow M \longrightarrow \quad \longleftarrow N \longrightarrow$

--- | A | 1 | 1 | 1 | --- | 1 | B | 1 | 1 | 1 | --- | 1 | A | ---

q_0

$Q = \{q_0, q_1, q_2, q_Y, q_N\}$, $F = \{q_Y, q_N\}$, $T = \{A, B, 0, 1, 2\}$

$(q_0, A, q_2, A, +1)$	$(q_1, A, q_N, *, *)$	$(q_2, A, q_Y, *, *)$
$(q_0, B, q_0, B, -1)$	$(q_1, B, q_1, B, +1)$	$(q_2, B, q_2, B, +1)$
$(q_0, 0, q_0, 0, -1)$	$(q_1, 0, q_1, 0, +1)$	$(q_2, 0, q_2, 0, +1)$
$(q_0, 1, q_1, 2, +1)$	$(q_1, 1, q_0, 0, -1)$	$(q_2, 1, q_0, 1, -1)$
$(q_0, 2, q_0, 2, -1)$	$(q_1, 2, q_1, 2, +1)$	$(q_2, 2, q_2, 1, +1)$

Suppose that we want to know whether an integer $M(>1)$ exactly divides a second integer N. Figure 8.1 shows a *DTM* which solves this decision problem. The upper half of the diagram shows the tape at the outset of the computation. Both M and N are required to be in unary representation for the algorithm (that is, the *TM*) to work. The tape symbols A and B are simply used as punctuation. We presume cell $C(0)$ is that occupied by the symbol B. The final state q_Y indicates a 'yes' answer while q_N indicates 'no'. The set of 15 quintuples listed in the diagram constitute the *TM*'s program, which operates by repeatedly subtracting M from N. If, whilst partially through one of these subtractions, none of N remains, then the *TM* is in state q_1 and the right-hand cell containing A is scanned. The appropriate quintuple then requires that the state q_N is entered. Notice that in this quintuple the character * indicates the irrelevance of specifying t_f or m. Similarly, if one of the subtractions of M from N has been completed and none of N remains, then the *TM* is in state q_2 and the right-hand A is scanned. The appropriate quintuple then causes the state q_Y to be entered.

For convenience, the theory of *NP*-completeness has been designed to apply only to decision problems. This is not restrictive because we can cast any problem into a closely related decision problem. For example, for the travelling salesman problem we might utilise the decision problem: does the graph G have a tour of length less than k? Here G is said to be an *instance* of the (travelling salesman) problem.

We are now in a position to define the class of problems called *P*. The class *P* consists of those decision problems for which there exists a polynomial time algorithm. In other words, *P* contains all those decision problems which can be solved, within $p(S)$ computational steps, by a *DTM* which halts for any input. Here $p(S)$ is a polynomial in S, the size of an instance of the problem. Thus *P* consists of those problems which can be efficiently solved.

Before introducing the class of problems called *NP* in a formal way, we shall try to capture the idea in an informal manner. For this purpose we refer to the travelling salesman decision problem defined earlier. In Chapter 6 we stated that there is no known efficient algorithm for the travelling salesman problem. However, consider the claim that the answer to an instance of the decision problem was 'yes' and that a particular tour was offered as evidence. It would be an easy matter to check this claim by determining whether or not the evidence indeed represented a tour and was of length less than K. Also, it is a simple matter to construct a polynomial time algorithm for this verification process. Of course, any set of edges may be a candidate for such a claim, and if it were possible to apply the verification algorithm *simultaneously* to all these sets then our decision problem could be solved in polynomial time. There is an extremely large class of decision problems, many of which have great practical importance, for which there is no known efficient algorithm, but yet which, like the travelling salesman problem, are *polynomial time verifiable*. It is this idea of polynomial time verifiability that the class of problems called *NP* is intended to capture.

We need to be certain that our terminology is clear. Given a verification algorithm (i.e., a *TM*), its input would consist of an *instance* of the decision problem and a *guess* for that instance. So, for example, in the case of the travelling salesman problem: does G contain a tour of length less than K?, any set of edges might be a guess and G is an instance of the problem.

We shortly describe a non-deterministic algorithm of the type we shall formally use to define the class *NP*. Such an algorithm will consist of two stages. The first stage simply produces a guess, placing this in the tape cells $C(-1)$, $C(-2)$, ..., $C(-l)$, and leaves the tape head over $C(0)$ in readiness for the second stage. An instance of the problem is already presumed to occupy the tape cells $C(1)$, $C(2)$, ..., $C(n)$. Any string of symbols from the tape alphabet will suffice for a guess. The guessing stage produces an arbitrary string by operating non-deterministically. That is, for any pair (q_i, t_i) there is (possibly) more than one triplet (q_f, t_f, m) and an arbitrary choice is made as to which one to apply when (q_i, t_i) arises. Q might contain two special guessing states q_a and q_b. The state q_a is a left moving state in that if the tape head is scanning a blank cell, then an arbitrary symbol from the tape alphabet is printed and if the new state is q_a then the tape head moves to the left. If, however, the new state is q_b then right moving results and thereafter the *TM* remains in this state, causing the tape head to move rightwards (leaving the content of tape cells unchanged) until $C(0)$ is encountered (which might be distinguished by containing a particular type symbol). Then the *TM* is made to enter state q_0 in readiness for the second state of the computation. This stage operates deterministically and attempts to verify the guess for the instance of the problem encoded within the tape cells $C(1)$ to $C(n)$.

As an example of such a non-deterministic algorithm or *TM* (*NDTM*) consider the decision problem: is N divisible by *some* $M(>1)$? For the second stage of operation we can make use of our previous example of a *TM* illustrated in figure 8.1. However, for the guessing stage we need some further quintuples and we also need to respecify the initial tape layout. This is illustrated in figure 8.2. The non-determinism here is provided by the first two of the additional quintuples listed in that diagram. This example is particularly simple because the guess is only constructed from one symbol, M being in unary representation. If a different encoding scheme were used, much more non-determinism would have to be built in.

For the example *NDTM* and indeed for any *NDTM* of the generic type we are describing there are several possible outcomes. The *NDTM* might halt in state q_Y, it might halt in state q_N or might not stop at all. We say

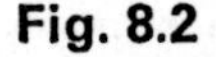

Fig. 8.2

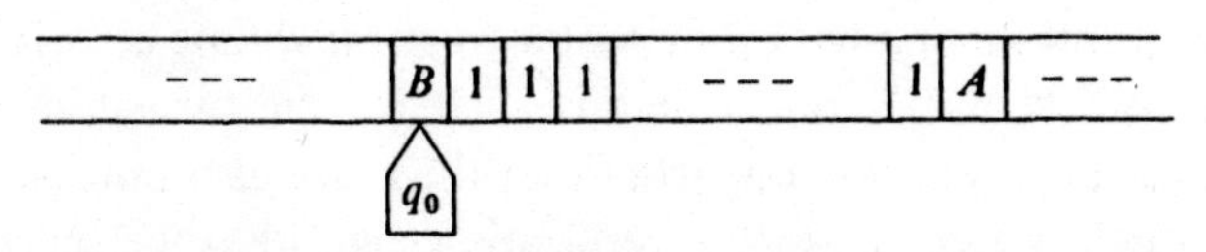

The program consists of the quintuples of figure 8.1 plus the following

$$(q_a, \text{blank}, q_a, 1, -1)$$
$$(q_a, \text{blank}, q_b, A, +1)$$
$$(q_b, \quad 1, q_b, 1, +1)$$
$$(q_b, \quad B, q_0, B, -1)$$
$$(q_0, \text{blank}, q_c, 1, -1)$$
$$(q_c, \text{blank}, q_a, 1, -1)$$
$$Q = \{q_0, q_1, q_2, q_Y, q_N, q_a, q_b, q_c\}$$

F and T are as shown in figure 8.1.

that an *NDTM solves* a decision problem D, if the following two conditions hold for all instances I of D:

(*a*) If D is true for I then there exists some guess for I which will lead to the *NDTM* to stop in state q_Y.

(*b*) If D is false for I then *no* guess exists for I which will lead to the *NDTM* to stop in state q_Y.

We have described an *NDTM* without reference to time-complexity. The informal preamble, however, made it clear that we are interested in polynomially bounded verification. Let us then define a polynomial time *NDTM*. The *NDTM* for the decision problem D is said to operate in polynomial time if for every instance I of D that is true, there is a guess that leads to the checking (or verification) stage of the *NDTM* to stop in state q_y within $p(S)$ computational steps. Here $p(S)$ is a polynomial in S, the length of I. An *NDTM* which operates in polynomial time is naturally called a *polynomial time NDTM*.

We are now in a position to define the class *NP*. *NP* is the class of all decision problems that can be solved by a polynomial time *NDTM*. *NP* is an acronym for non-deterministically polynomially bounded.

Clearly, any problem $D \in P$ is also contained in *NP*, because we can obtain a polynomial time *NDTM* for D by using the polynomial time *DTM* for D as the checking stage, and this machine will respond with 'yes' whilst ignoring guesses. It is an outstanding and important question in complexity theory as to whether or not $P = NP$. There is a widespread belief, based on a great deal of circumstantial evidence, but not on a proof, that *NP* is a much larger class than *P*.

We finish this section by noting that a *polynomial* time algorithm is not precisely defined unless the definition includes the format of its input (exercise 8.7 illustrates the importance of this). We also note that for the execution of a polynomial time verification algorithm we can disregard any guess longer than $p(n)$ – the polynomial bounding the computation.

8.1.2. NP-completeness and Cook's theorem

The unresolved question as to whether or not $P = NP$ is important because if $P \neq NP$ then whilst the problems in *P* can be efficiently solved, those in $(NP-P)$ would be intractable. It seems that a resolution of this fundamental question will be difficult to obtain so that the theory of *NP*-completeness, which we introduce in this section, concentrates on the weaker question: if $P \neq NP$ then does the problem in hand belong to $(NP-P)$?

A basic idea in the theory of *NP*-completeness is that of a *polynomial transformation.* Let D_1 and D_2 denote two decision problems. We say that there is a polynomial transformation from D_1 to D_2, written $D_1 \propto D_2$, if the following two conditions hold:

(*a*) There exists a function $F(I)$ transforming any instance I of D_1 to an instance of D_2 such that the answer to I with respect to D_1 is 'yes' if and only if the answer to $F(I)$ is 'yes' with respect to D_2.
(*b*) There exists an efficient algorithm to compute $F(I)$.

Notice that the *TM* implied by (*b*) would not supply a 'yes' or 'no' answer but would print $F(I)$ on the tape given I as input.

If $D_1 \propto D_2$ and if there is a polynomial time algorithm for D_2, then there is a polynomial time algorithm for D_1. We denote the polynomial bounding the computation of $F(I_1)$ by $p_1(S_1)$ where S_1 is the length of I_1, and the polynomial bounding the computation of D_2 is denoted by $p_2(S_2)$ where S_2 is the length of $I_2 = F(I_1)$. Clearly, $S_2 \leqslant p_1(S_1)$ and so the computation time for D_1, consisting of a transformation to D_2 followed by the computation of D_2, is bounded by $p_1(S_1)+p_2(p_1(S_1))$ which is clearly a polynomial in S_1. It is easy to see that if $D_1 \propto D_2$ and $D_2 \propto D_3$ then $D_1 \propto D_3$.

We are now in a position to define *NP*-complete problems. A decision problem, D, is said to be *NP*-complete, written *NPC*, if $D \in NP$ and if for every problem $D' \in NP$, $D' \propto D$. The importance of this definition can be seen from the following two observations:

(*a*) If D is *NPC* and if $D \in P$ then $NP = P$.
(*b*) If D is *NPC*, $D \propto D'$ and $D' \in NP$ then D' is *NPC*.

Thus the set of problems which are *NPC* form an equivalence class of what might be considered the most difficult problems in *NP*. The next theorem, a celebrated one due to Cook, provides us with the first problem that was known to be *NPC*. Using this basis along with the fact that the relation of polynomial transformation is transitive, hundreds of problems have been shown to be *NPC*. Many of these have important applications and many are to be found in graph theory.

Before describing Cook's theorem we need to define the problem of the *satisfiability of conjunctive normal forms*, or *SAT* for short. Given a finite set $V = \{v_1, v_2, ..., v_n\}$ of logical variables, we define a *literal* to be a variable v_i or its complement, $\bar{v}_i$. If $v_i =$ **true** then $\bar{v}_i =$ false and vice-versa. We define a *clause*, C_i, to be a set of literals. An instance, I, of *SAT*, consists of a set of clauses (any literal may appear in any number of these clauses). The problem of *SAT* is whether or not there exists a *truth assignment* (i.e., an assignment of the values **true** or **false** to each member of V) such that at least one member of each clause of I has the value **true**. If the answer is 'yes', then we say that I has been *satisfied*. Let us restate the problem in a format which from now on we take to be a standard means to describe decision problems.

SAT:

Instance: A set of clauses, $\{C_i\}$, over the set of logical variables, V.
Question: Is there a truth assignment to V such that $\{C_i\}$ is satisfied?

Theorem 8.1 (Cook). *SAT* is *NPC*.

Proof. A non-deterministic algorithm for *SAT* has simply to check that any truth assignment satisfies each clause in an instance of the problem. It is a simple matter to construct a polynomial time *NDTM* to do this. Thus $SAT \in NP$.

The more complex part of this proof is to show that every problem in *NP* is polynomially transformable to *SAT*. In order to do this we shall construct a mapping from any instance I of an *arbitrary* polynomial time *NDTM*, M, to instances $F(I)$ of *SAT*. This will be done in such a way that $F(I)$ is satisfiable if and only if M responds with 'yes' for some guess applied to I. As we shall see, the idea behind $F(I)$ is a simulation of M for the instance I.

Let us define the set of states Q of M, and the set of tape symbols T, as follows:

$$Q = \{q_0, q_1, ..., q_r\} \text{ where } q_1 = q_Y \text{ and } q_2 = q_N$$

$$T = \{t_0, t_1, ..., t_s\} \text{ where } t_0 = \text{blank}$$

and we take I, contained in the tape cells $C(1)$ to $C(n)$ to be:

$$I = t_{k_1}, t_{k_2}, \ldots, t_{k_n}$$

In order to construct $F(I)$ we need to define a set of logical variables. This set will be a union of three subsets:

$$\{Q(i, k)\} \cup \{H(i,j)\} \cup \{S(i,j,l)\}$$

The interpretation to be placed on each variable is as follows: at time i, $Q(i, k)$ specifies as **true** or **false** that M is in state q_k, $H(i,j)$ specifies that the tape head of M is scanning $C(j)$ while $S(i,j,l)$ specifies that the content of $C(j)$ is t_l. M is a polynomial time *NDTM* so that i and j are bounded as follows:

$$0 \leqslant i < p(n), \; -p(n) \leqslant j \leqslant p(n)+1$$

where $p(n)$ is a polynomial in the length n of I. Of course, both Q and T are finite:

$$0 \leqslant k \leqslant r, 0 \leqslant l \leqslant s$$

where r and s depend on M. It follows that the number of H-variables and the number of S-variables are of order $(p(n))^2$ while the number of Q-variables is of order $p(n)$.

Clearly, a truth assignment to the logical variables we have defined might correspond to a computation of M, although an arbitrary assignment will probably not. Indeed such an assignment may imply that M is simultaneously in several states and that several tape cells are being scanned. The construction of $F(I)$ which we now describe, builds a set of clauses, each of which is designed to ensure one requirement that $F(I)$ models a computation on M. $F(I)$ contains six groups of clauses as follows: (how many clauses are in each group is determined by i, j, k and l, where they occur, taking on all possible permutations of values consistent with the ranges shown above):

(*a*) At the outset of a computation M is in state q_0, cell $C(0)$ is being scanned and I is contained in the cells $C(1)$ to $C(n)$. We shall also assume the convention that when $i = 0$, $C(0)$ and $C(n+1)$ to $C(p(n)+1)$ are blank. (This convention was not used in our earlier example when I was delimited with the punctuation symbols A and B. The use of punctuation symbols can greatly reduce the number of states required by a *TM*, but they can be avoided. See for example exercise 8.5(*b*).) These observations result in the following clauses for $F(I)$:

(*a*) $\{Q(0, 0)\}, \{H(0, 1)\}, \{S(0, 0, 0)\}$
$\{S(0, 1, k_1)\}, \{S(0, 2, k_2)\} --- \{S(0, n, k_n)\}$
$\{S(0, n+1)\}, \{S(0, n+2, 0)\} --- \{S(0, p(n)+1, 0)\}$

(*b*) At any time M is in at least one state:
$\{Q(i,0), Q(i,1) --- Q(i,r)\}$
but in not more than one state:
$\{\overline{Q(i,j)}, \overline{Q(i,j')}\}, 0 \leqslant j < j' \leqslant r$

(*c*) At any time exactly one tape cell is being scanned (cf. (*b*))
$\{H(i, -p(n)), H(i, -p(n)+1), --- H(i, p(n)+1)\}$
$\{\overline{H(i,j)}, \overline{H(i,j')}\}, -p(n) \leqslant j < j' \leqslant p(n)+1$

(*d*) At any time each tape cell contains exactly one symbol (cf. (*b*)):
$\{S(i,j,0), S(i,j,1), --- S(i,j,s)\}$
$\{\overline{S(i,j,l)}\ \overline{S(i,j,l')}\}, 0 \leqslant l < l' \leqslant s$

(*e*) By the time $i = p(n)$, M has entered state q_Y:
$\{Q(p(n), 1)\}$

(*f*) The changes in M from one computational step to the next are dictated by a quintuple of M and so $F(I)$ includes:
$\{\overline{H(i,j)}, \overline{Q(i,k)}, \overline{S(i,j,l)}, H(i+1, j+m)\}$
$\{\overline{H(i,j)}, \overline{Q(i,k)}, \overline{S(i,j,l)}, Q(i+1, k')\}$
$\{\overline{H(i,j)}, \overline{Q(i,k)}, \overline{S(i,j,l)}, S(i+1, l')\}$
where if $q_k \in Q-\{q_Y, q_N\}$ then the values of l', k' and m are provided by the quintuple $(q_k, t_l, q_{k'}, t_{l'}, m)$ whilst if $q_Y \in \{q_Y, q_N\}$ then $k' = k$, $l' = l$ and $m = 0$.

The set of clauses which is a union of those described in (*a*) to (*f*) constitutes an instance of *SAT*. As we have carefully explained, the construction ensures that $F(I)$ is satisfied if and only if the truth assignment describes a computation of M which halts in state q_Y.

It is easy to see that $F(I)$ can be constructed in polynomial time. As we explained earlier, the fact that M is polynomially bounded means that the number of variables of $F(I)$ is polynomially bounded. This in turn leads to a polynomial bound on the number of clauses constructed in (*a*) to (*f*). For example, from (*f*) we obtain $6\,p(n)\,(p(n)+1)\,[(r+1)\,(s+1)]$ clauses, where r and s are fixed by M.

Thus *SAT*, by all requirements, is *NPC*. ■

For the purposes of proving that other problems are *NPC* it is often simpler to transform to a subproblem of *SAT*, namely 3*SAT*, which we now define and prove to be *NPC*.

3SAT

Instance: A set of clauses $\{C_i\}$ each clause containing precisely three literals, over the set of logical variables, V.
Question: Is there a truth assignment to V such that $\{C_i\}$ is satisfied?

Theorem 8.2. $3SAT$ is NPC.

Proof. Clearly, $3SAT \in NP$ just as SAT is. In order to complete the proof we show that $SAT \propto 3SAT$. Let $C = \{a_1, a_2, ..., a_l\}$ be any one of the clauses in an instance of SAT. We shall show that C can be replaced by a number of clauses, each containing three literals, in such a way that these clauses are satisfied if and only if C is. We denote by C' the set of clauses replacing C. C' will utilise a number, depending on l, of dummy variables. We denote by V_c this set of introduced variables. There are a number of cases depending upon l:

(*a*) $l = 1, V_c = \{x_1, x_2\}$
$C' = \{\{a_1, x_1, x_2\}, \{a_1, x_1, \bar{x}_2\}, \{a_1, \bar{x}_1, x_2\}, \{a_1, \bar{x}_1, \bar{x}_2\}\}$
(*b*) $l = 2, V_c = \{x_1\}$
$C' = \{\{a_1, a_2, x_1\}, \{a_1, a_2, \bar{x}_1\}\}$
(*c*) $l = 3, V_c = \varnothing$
$C' = \{C\}$
(*d*) $l > 3, V_c = \{x_i | 1 \leqslant i \leqslant l-3\}$
$C' = \{\{a_1, a_2, x_1\}\} \cup \{\{\bar{x}_i, a_{i+2}, x_{i+1} | 1 \leqslant i \leqslant l-4\}\}$
$\cup \{\{\bar{x}_{l-3}, a_{l-1}, a_l\}\}$

Let us consider these cases in turn. A truth assignment satisfying C under the cases (*a*) and (*b*) will clearly satisfy each clause in C' whatever assignment is made to the dummy variable(s). Moreover, if C' is satisfied then so is C. In these cases we could arbitrarily assign the value **true** to each member of V_c. Case (*c*) is trivially alright.

The outstanding case, (*d*), is a little more involved. If there is a satisfying truth assignment for C then at least one literal in C is **true**. Let a_p be such a literal. If $p = 1$ or 2, then we let $x_i =$ **false** for $1 \leqslant i \leqslant l-3$, if $p = l$ or $(l-1)$ then we let $x_i =$ **true** for $1 \leqslant i \leqslant l-3$, otherwise we let $x_i =$ **true** for $1 \leqslant i \leqslant p-2$ and $x_i =$ **false** for $p-1 \leqslant i \leqslant l-3$. It is easy to see that these assignments will cause C' to be satisfied if and only if C is.

Hence any instance of SAT is transformable to an instance of $3SAT$. Moreover, this transformation can be achieved in polynomial time. In order to see this we just need observe that if an instance of SAT has m clauses of maximum length l, then the corresponding instance of $3SAT$ has at most lm clauses. Thus $3SAT$ is NPC. ∎

We are now in a position to apply this introduction to the theory of NP-completeness to some of the problems of graph theory. The remaining part of this chapter is devoted to that end.

8.2 *NP*-complete graph problems

In this section we show that several important problems in graph theory are *NPC*. A great number of problems in graph theory are *NPC* so that our selection is a relatively small one. The interested reader might consult the long list of *NPC* graph problems in Garey & Johnson.[1]

The proof that a decision problem D is *NPC* would normally consist of two steps:

(*a*) that $D \in NP$,

and

(*b*) that $D' \propto D$ for some problem D' that is *NPC*.

In all the proofs that follow (*a*) is relatively trivial so that we often adopt the practice of omitting that step. The proofs shall therefore concentrate on the more difficult step, (*b*). At present the only candidates for D' are *SAT* or 3*SAT*. As we proceed we expand our choice in this respect, building up transformation chains of *NPC* problems.

We divide this section into a number of subsections each consisting of a selection of closely related problems. Throughout G is a simple graph with a vertex set V, where $|V| = n$.

8.2.1. Problems of vertex-cover, independent set and clique

A *vertex-cover* of G consists of a subset $V' \subseteq V$ such that for every edge (u, v) of G, at least one of u and v is in V'. The *size* of a vertex-cover is given by $|V'|$. We show that the following problem is *NPC*:

Vertex cover (*VC*):

Instance: A graph G and an integer k, $1 \leqslant k \leqslant n$.
Question: Does G have a vertex-cover V' such that $|V'| \leqslant k$?

Theorem 8.3. *VC* is *NPC*.

Proof. We show that 3*SAT* $\propto$ *VC*. Given an instance of 3*SAT* consisting of a set of clauses $C = \{c_1, c_2, \ldots, c_i\}$ over the set of variables

$$U = \{u_1, u_2, \ldots, u_j\}$$

we construct an instance of *VC* consisting of the graph G and an integer k as follows. For every variable $u_r \in U$, G contains two vertices v_r and $\bar{v}_r$, representing u_r and its complement, and the edge $(v_r, \bar{v}_r)$. For every clause $c_s = \{l_1, l_2, l_3\} \in C$, G contains the vertices v_1^s, v_2^s and v_3^s, and the edges (v_1^s, v_2^s), (v_2^s, v_3^s) and (v_1^s, v_3^s). The construction is completed by adding, for each clause c_s, the edges (v_1^s, l_1), (v_2^s, l_2) and (v_3^s, l_3); finally we set $k = j+2i$.

Figure 8.3 shows an instance of VC obtained by this construction from this set of clauses:

$$\{\{u_1, u_2, u_3\}, \{\bar{u}_1, \bar{u}_3, u_4\}, \{u_2, u_3, \bar{u}_4\}\}$$

In this case, $k = 10$. Since G has $(2j+3i)$ vertices and $(j+6i)$ edges it follows that our construction can be achieved in polynomial time. In order to complete the proof we just need show that C is satisfiable if and only if G has a vertex-cover of size k or less.

Fig. 8.3

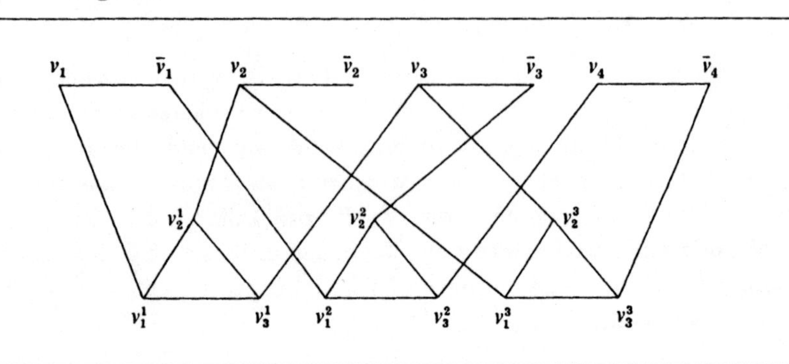

We first notice that any subset of vertices of G that is a vertex-cover, must include at least one of the vertices v_r and $\bar{v}_r$ for each r, $1 \leqslant r \leqslant j$, and at least two of the vertices v_1^s, v_2^s and v_3^s for each s, $1 \leqslant s \leqslant i$. In other words, a vertex-cover contains *at least* $j+2i$ vertices. Thus the instance of VC constructed according to our rules can only be true if V' contains *exactly* one vertex from each pair, $(v_r, \bar{v}_r)$ and *exactly* two vertices from each triple (v_1^s, v_2^s, v_3^s), because $k = j+2i$. Suppose V' is a vertex-cover of G. A satisfactory truth assignment for C can then be obtained by setting $u_m =$ **true** if the vertex labelled v_m is in V', otherwise we set $u_m =$ **false**. That C is satisfied can be seen as follows. Consider the three edges from the vertex-set $\{v_1^s, v_2^s, v_3^s\}$ to those vertices representing the literals of $3SAT$. Precisely two of these edges will be covered by vertices in $\{v_1^s, v_2^s, v_3^s\} \cap V'$, this means that the third must be covered by a vertex representing a literal in $c_s = \{l_1, l_2, l_3\}$. According to our truth assignment this literal is **true** and so c_s is satisfied. We can apply the same argument to every clause in C and so it follows that if a vertex-cover of size k exists for G then there is a satisfying truth assignment for C.

Conversely, let T be a truth assignment satisfying C. We can construct a vertex-cover V' for G, of size k, as follows. For each pair of vertices labelled v_m and $\bar{v}_m$, V' includes v_m if T assigns the value **true** to u_m, otherwise

V' includes $\bar{v}_m$. This ensures that at least one of the three edges from each vertex-set $\{v_1^s, v_2^s, v_3^s\}$ to those vertices representing the literals of $3SAT$ are covered. We can cover the other two by including their end-points in V' which are in $\{v_1^s, v_2^s, v_3^s\}$. This clearly provides a vertex-cover. ∎

The following two problems are very strongly related to VC so that proofs of their NP-completeness easily follows.

Independent set (IS):

Instance: A graph G and an integer k, $1 \leqslant k \leqslant n$.

Question: Does G contain an independent set of size greater than or equal to k?

Theorem 8.4. *IS* is *NPC.*

Proof. We show that $VC \propto IS$. Given an instance of VC consisting of G and the integer k', we can construct an instance of IS within polynomial time which consists of G and $k = n - k'$. Clearly, V' is a vertex-cover of G if and only if $(V - V')$ is an independent set of G. It immediately follows that $VC \propto IS$ and that IS is NPC. ∎

CLIQUE:

Instance: A graph G and an integer k, $1 \leqslant k \leqslant n$.

Question: Does G contain a clique of size greater than or equal to k?

Theorem 8.5. *CLIQUE* is *NPC.*

Proof. We show that $IS \propto CLIQUE$. Given an instance of IS consisting of G and k, we can construct an instance of $CLIQUE$ consisting of G', the complement of G with respect to the complete graph on n vertices, and k. Clearly, G' can be constructed within polynomial time. Now $V' \subseteq V$, is an independent set of G if and only if the vertices of V' form a clique in G'. Thus $IS \propto CLIQUE$ and so $CLIQUE$ is NPC. ∎

8.2.2. Problems of Hamiltonian paths and circuits and the travelling salesman problem

The following set of problems, like the previous set, are strongly related, so that having proved the first to be NPC, proofs for the others follow easily. Our first relatively difficult proof concerns the existence of a Hamiltonian path in a directed graph:

Directed Hamiltonian path (DHP)

Instance: A directed graph G with two distinguished vertices, u and v.

Question: Does G contain a directed Hamiltonian path from u to v?

Theorem 8.6. *DHP* is *NPC*.

Proof. We show that $VC \propto DHP$. We first describe the construction of a specific instance of *DHP* from a specific instance of *VC* with the use of figure 8.4. As we shall see, this construction is easily generalised. Figure 8.4(a) shows an instance of *VC* comprised of the graph G and the integer $k = 2$. Each edge (u, v) of G has two identifiers, $e(u, l)$ and $e(v, m)$ signifying that (u, v) is the lth edge incident with u and the mth edge incident with v. We now construct the instance of *DHP* consisting of the directed graph G' and two distinguished vertices u_0 and u_k. This is done in three stages as follows.

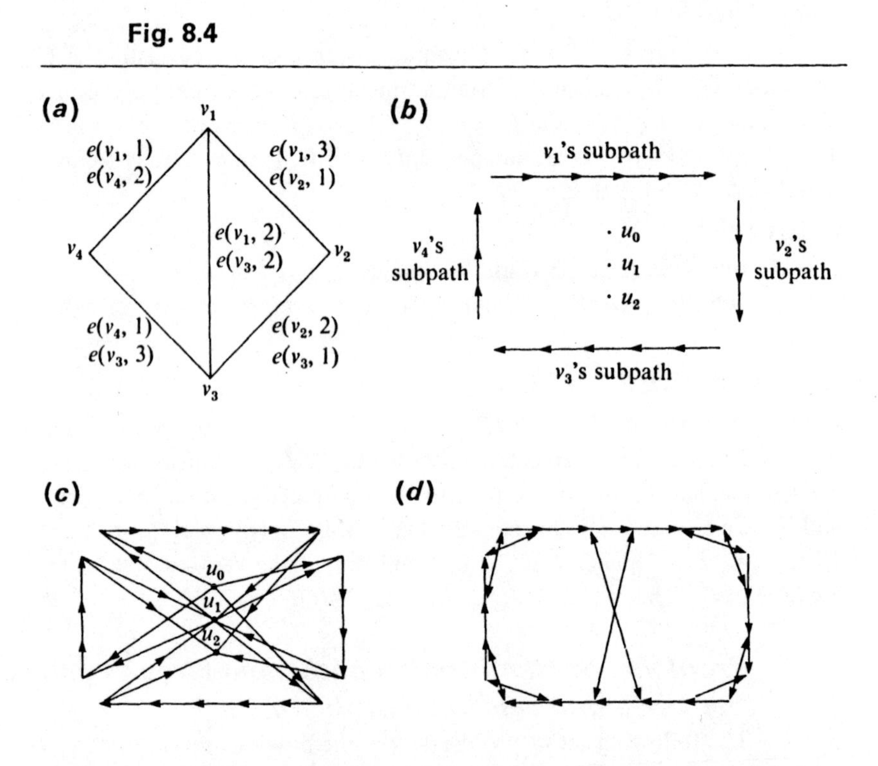

Fig. 8.4

For each vertex v of G, G' contains $2 \times d(v)$ vertices where $d(v)$ is the degree of v. These vertices are linked by a directed path, called v's subpath. To complete the first stage of our construction, which is shown in figure 8.4(b) we add to G', $(k+1)$ vertices: u_0, u_1 and u_2. The second stage of the construction consists of adding edges from u_0 and u_1 (that is, from u_0, u_1, ..., u_{k-1}) to the tail of each v's subpath and adding arcs from the head of each v's subpath to u_1 and u_2 (that is, to u_1, u_2, ..., u_k). At this stage the construction of G' is shown in figure 8.4(c). Finally, if $e(u, l)$ and $e(v, m)$ identify the same edge in G, then the head(tail) of the $(2l-1)$th arc in u's subpath is linked in both directions to the head(tail) of the $(2m-1)$th arc in v's subpath. Figure 8.4(d) shows these additional arcs, where for clarity, we have omitted those arcs introduced in the second stage of our construction and the vertices $u_0, u_1, \ldots, u_k$. Again, for clarity, two oppositely directed arcs between the same pair of vertices have been merged into a single line. Using this specific example as a guide we can now describe the construction for an arbitrary case.

Let G and k denote an instance of *VC* and G' be the graph in an instance of *DHP* constructed from it. For every vertex v of G there are $2 \times d(v)$ vertices in G', each denoted by a triple: $(v, i, 1)$ and $(v, i, 2)$ for all i, $1 \leqslant i \leqslant d(v)$. There is a directed path through each such set of vertices, called v's subpath, consisting of the following edges (directed from the first to the second vertex):

$$((v, i, 1), (v, i, 2)) \quad \text{and} \quad ((v, i, 2), (v, i+1, 1))$$
$$\text{for all } i, 1 \leqslant i < d(v)$$

G' also contains a set of vertices, $\{u_0, u_1, \ldots, u_k\}$ and the edges:

$$(u_i, (v, 1, 1)) \quad \text{for all } i, 0 \leqslant i < k$$
$$((v, d(v), 2), u_i) \quad \text{for all } i, 0 < i \leqslant k$$

Finally, G' also contains the edges:

$$((u, i, 1), (v, j, 1)) \quad \text{and} \quad ((u, i, 2), (v, j, 2))$$

for every edge in G which is identified by $e(u, i)$ and $e(v, j)$. In order to complete the instance of *DHP* we simply specify u_0 to be the initial vertex and u_k the final vertex of the proposed path. It is easy to see that the number of vertices and the number of edges in G' are both bound by a polynomial in n, the number of vertices in G. Hence the construction can be achieved in polynomial time.

In order to complete the proof we need to show that G has a vertex-cover of size k or less, if and only if there is a Hamiltonian path from u_0 to u_k in G'.

Let us first suppose that G has a vertex-cover of size $\leqslant k$, then it must have a vertex-cover of size k. We denote such a cover by $C = \{v_1, v_2, \ldots, v_k\}$. A directed Hamiltonian path from u_0 to u_k in G' can then be constructed as follows. The first edge in the path is $(u_0, (v_1, 1, 1))$ followed by v_1's subpath and then the edge $((v_1, d(v_1), 2), u_1)$; we then similarly pass from u_1 to u_2 via v_2's subpath and so on until vertex u_k is reached. The path from u_0 to u_k we have described does not as yet include those vertices on any v_l

subpath where $l > k$. Suppose that (u, v) is an edge in G identified by $e(u, i)$ and $e(v, j)$, and that $u \notin C$. We can include $(u, i, 1)$ and $(u, i, 2)$ in the path by making a detour from v's subpath as follows: replace the edge $((v, j, 1), (v, j, 2))$ by the sequence $((v, j, 1), (u, i, 1))$, $((u, i, 1), (u, i, 2))$ and $((u, i, 2), (v, j, 2))$. Every vertex on the unused v_l subpaths can be included in this way because our construction of G' ensures that the appropriate edges are present given that C is a vertex-cover.

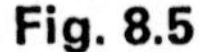

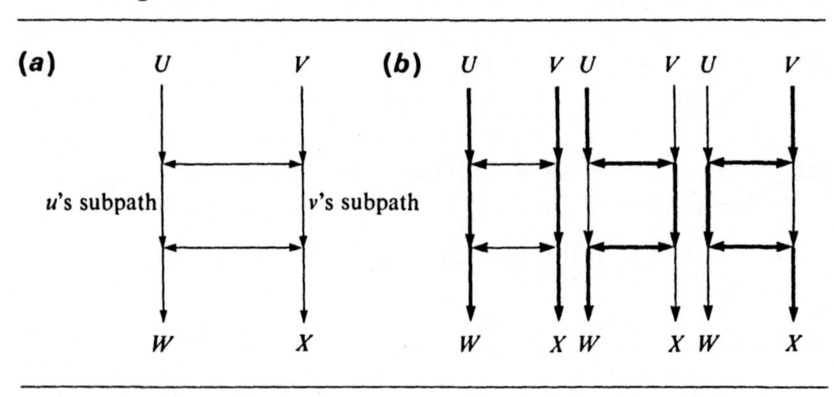

Conversely, suppose that G' has a directed Hamiltonian path from u_0 to u_k. Suppose that (u, v) is an edge in G identified by $e(u, i)$ and $e(u, j)$ and consider the vertices $(u, i, 1)$, $(u, i, 2)$, $(v, j, 1)$ and $(v, j, 2)$ in G'. These are shown in figure 8.5(*a*). A Hamiltonian path passing through these vertices can only approach them from U and/or V. In order that all these vertices are included in such a path only three routes are possible for it, these are shown as the heavily scored paths in figure 8.5(*b*). Thus a Hamiltonian path entering from U must exit at W, and one entering from V must exit at X. This means that if $(v, 1, 1)$ is approached from some u_i on a directed Hamiltonian path, then every vertex on v's subpath is visited before another (different) u_i is visited by traversing the edge $((v, d(v), 2), u_i))$. In this circumstance the Hamiltonian path is said to use v's subpath. Given our supposition that G' contains a Hamiltonian path, H, we construct a vertex-cover C, of size k for G, by including in C all those vertices whose subpaths are used in H. In order that all the vertices $(u, i, 1)$, $(u, i, 2)$, $(v, j, 1)$ and $(v, j, 2)$ are included in H at least one of the subpaths belonging to u and v must be used, and so the edge $(u, v) \in G$ is covered by this construction of C. This concludes our proof. ■

Having proved that *DHP* is *NPC* we are now in a position to provide quick proofs that the remaining problems in this section are also *NPC*.

Directed Hamiltonian Circuit (DHC)

Instance: A *directed* graph G.

Question: Does G contain a directed Hamiltonian circuit?

Theorem 8.7. *DHC is NPC.*

Proof. We can easily see that $DHP \propto DHC$ as follows. Given an instance of *DHP* consisting of a digraph G' and the vertices u and v, we construct an instance of *DHC*, G, by adding the edge (v, u) to G'. Obviously there is a directed Hamiltonian circuit in G if and only if there is a directed Hamiltonian path in G' from u to v. Thus $DHP \propto DHC$ and hence *DHC* is *NPC*. ∎

Hamiltonian Path (HP)

Instance: A graph G with two distinguished vertices u and v.

Question: Does G contain a Hamiltonian path between u and v?

Theorem 8.8. *HP is NPC.*

Proof. We shall show that $DHP \propto HP$. Let G' and the two vertices v_a and v_b be an instance of *DHP*. We construct an undirected graph G from G' as follows. For every vertex v_i of G', G contains three vertices v_i^1, v_i^2 and v_i^3, and the edges (v_i^1, v_i^2) and (v_i^2, v_i^3). For each edge (v_i, v_j) of G', G contains the edge (v_i^3, v_j^1). Our instance of *HP* then consists of G and the vertices v_a^1 and v_b^3. Figure 8.6 shows an instance of *HP* in (*b*) constructed in this way from the instance of *DHP* shown in (*a*):

Fig. 8.6

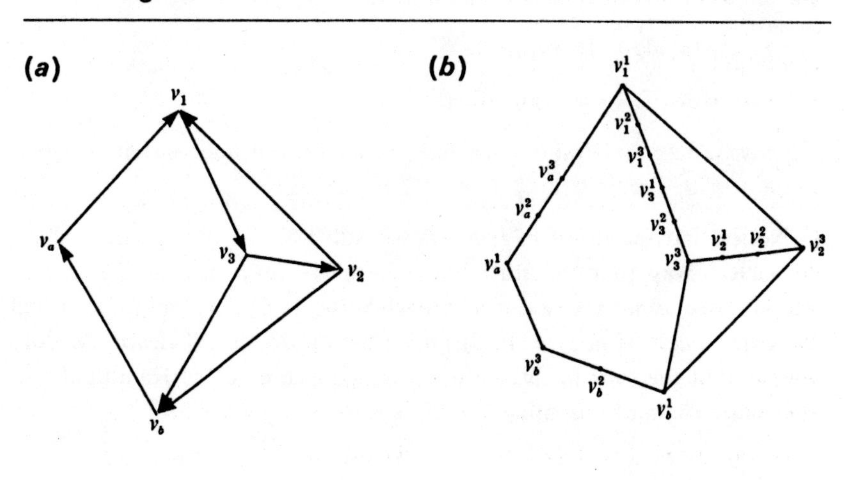

We complete the proof by showing that G has a Hamiltonian path between v_a^1 and v_b^3 if and only if G' has a directed Hamiltonian path from v_a to v_b.

Let G' have such a path, H'. G then has the Hamiltonian path consisting of the edges (v_k^1, v_k^2), (v_k^2, v_k^3), for all $v_k \in G'$, and the edges (v_i^3, v_j^1), for all $(v_i, v_j) \in H'$. Conversely suppose G has a Hamiltonian path, H. H must contain the edges (v_i^1, v_i^2) and (v_i^2, v_i^3), otherwise v_i^2 could not be reached. Moreover, if in following H from v_a^1 to v_b^3, we visit the vertices v_i^1, v_i^2 and v_i^3 in the order of writing them, then for any other vertex $v_j \in G'$, we visit v_j^1, v_j^2, v_j^3 in G in the order that we have written them. It follows that if $H = v_a^1, v_a^2, v_a^3, v_1^1, v_1^2, v_1^3, \ldots, v_b^1, v_b^2, v_b^3$, then G' contains a directed Hamiltonian path, $v_a, v_1, \ldots, v_b$. ∎

The question as to whether or not an undirected graph G contains a Hamiltonian circuit (*HC*) is, of course, easily shown to be *NPC*. A proof would simply show that $HP \propto HC$ rather like theorem 8.7 shows that $DHP \propto DHC$.

We come now to the final problem of this section.

Travelling salesman (TS)

Instance: A weighted complete graph G and an integer $k > 0$.
Question: Does G contain a Hamiltonian circuit of length $\leqslant k$?

Theorem 8.9. *TS* is *NPC*.

Proof. We show that $HC \propto TS$. Let G' be an instance of *HC* with n vertices. We construct an instance of *TS* consisting of a graph G and $k = n$ as follows. G is the complete graph on the vertices of G' with edge-weights $w(e)$, for each edge e, as follows:

$$w((u, v)) = 1 \quad \text{if } (u, v) \in G'$$

and

$$w((u, v)) = 2 \quad \text{if } (u, v) \notin G'$$

Clearly, G has a Hamiltonian circuit of length n if and only if G' has a Hamiltonian circuit. Thus $HC \propto TS$ and therefore *TS* is *NPC*. ∎

Notice that the proof of theorem 8.9 still holds if we had defined *TS* in such a way that the travelling salesman's tour was *not necessarily* Hamiltonian, but that nevertheless each vertex (i.e., city) had to be visited *at least* once. Of course, the transformation $HC \propto TS$ described does ensure that the tour in the constructed instance of *TS* is Hamiltonian if and only if it is of minimum length.

8.2.3. Problems concerning the colouring of graphs

The following problem is *NPC*:

k-colouring (KC)

Instance: A graph G and an integer k, $1 \leqslant k \leqslant n$.
Question: Does there exist a (proper, vertex-)colouring of G using $\leqslant k$ colours?

We shall first show that a restriction of the problem (in which $k = 3$) is *NPC*. Before doing so we note the following lemma:

Lemma 8.1. In a 3-colouring of the following graph:

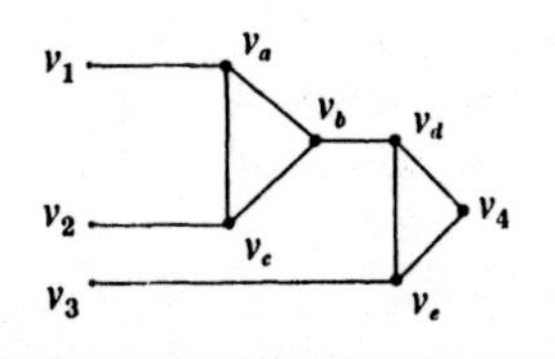

using the colours 0, 1 and 2, v_4 *must* be coloured 0 if and only if all the vertices v_1, v_2 and v_3 are coloured 0.

Proof. We leave this as an easy exercise for the reader. From now on we use the following shorthand to specify this graph:

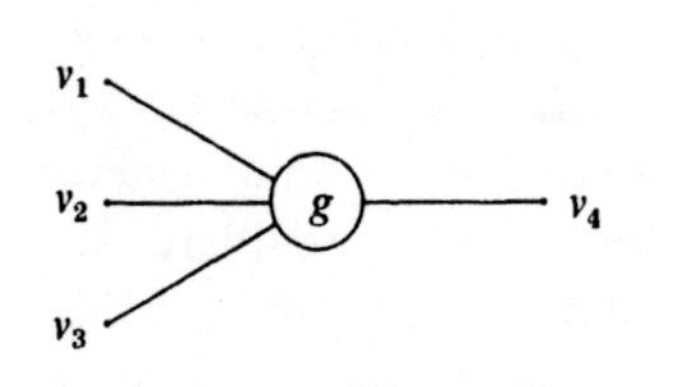

Theorem 8.10. $3C$ is *NPC*.

Proof. We show that $3SAT \propto 3C$. Given an instance of $3SAT$ consisting of the set of clauses $C = \{c_1, c_2, \ldots, c_p\}$ over the set of variables

$$U = \{u_1, u_2, \ldots, u_q\},$$

we construct an instance G of $3C$ as follows. G contains vertices labelled u_i and $\bar{u}_i$ for each $u_i \in U$, and the edges $(u_i, \bar{u}_i)$. For each clause,

$$c_j = \{l_1, l_2, l_3\} \in C,$$

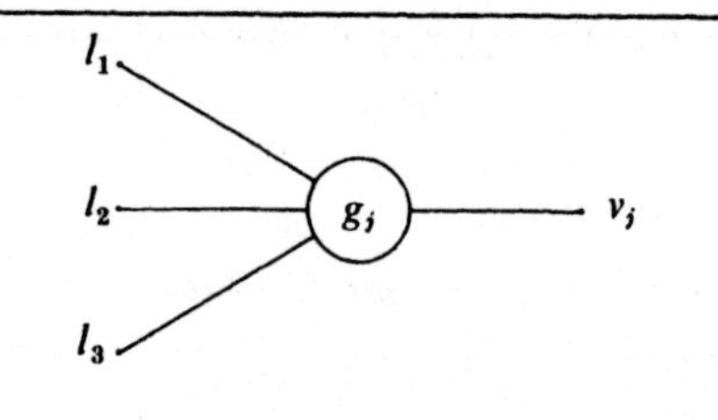

G contains a subgraph of the type specified in the above lemma: Notice that each of l_1, l_2 and l_3 is one of the vertices u_i or $\bar{u}_i$ for some i. G also contains a vertex labelled a and the edges (a, v_j) for all j, $1 \leqslant j \leqslant p$. Finally, G contains a vertex b with incident edges (b, u_i), $(b, \bar{u}_i)$ for all i, $1 \leqslant i \leqslant q$. Figure 8.7 shows this construction of G from the following instance of $3SAT$:

$$C^1 = \{(\bar{u}_1, u_2, u_3), (u_1, u_3, \bar{u}_4), (\bar{u}_2, u_3, u_4)\}$$

Fig. 8.7

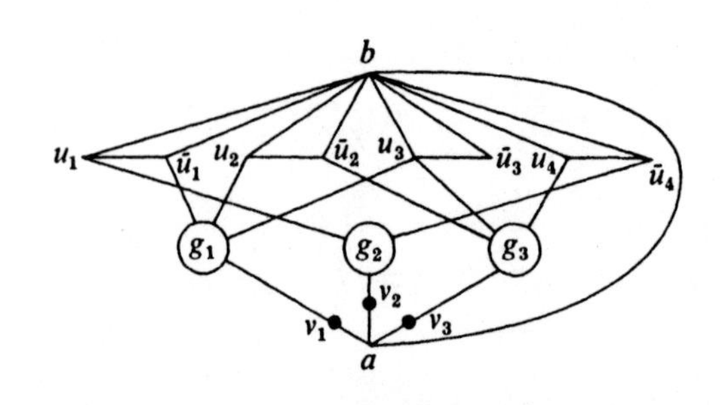

Suppose that C has a satisfying truth assignment, T. If u_i = **true** then the vertex $u_i \in G$ is coloured 1 and $\bar{u}_i$ is coloured 0, otherwise u_i is coloured 0 and $\bar{u}_i$ is coloured 1. Since T is a satisfying truth assignment, not all the vertices labelled l_1, l_2 and l_3 for the subgraph g_j of G, $1 \leqslant j \leqslant p$, are coloured 0. Thus by lemma 8.1, the vertices v_j of g_j can be labelled with 1 or with 2. Thus vertex a can be coloured 0. The only vertex as yet uncoloured is b. This is adjacent only to vertices coloured 0 or 1, and so we complete a 3-colouring by assigning the colour 2 to b.

Conversely suppose that G has a 3-colouring. Without loss of generality, we can assume that b is coloured 2 and that a is coloured 0. This implies that the vertices u_i and $\bar{u}_i$, $1 \leqslant i \leqslant q$, are coloured 0 or 1 and that the vertices v_j, $1 \leqslant j \leqslant p$, are coloured 1 or 2. By lemma 8.1, for each subgraph g_j of G, it cannot be that all of l_1, l_2 and l_3 are coloured 0. A satisfying truth assignment is then obtained by assigning the value **true** to a literal if and only if its corresponding vertex in G is coloured 1.

Thus $3SAT \propto 3C$ and hence $3C$ is NPC. ■

A proof that KC is NPC for any $K > 3$ can be established by a proof similar to but generally more tedious than that for theorem 8.10. The details include defining a special subgraph analogous to that described in lemma 8.1 (notice that this has two complete subgraphs on $K = 3$ vertices) and the replacement of the vertices a and b by a complete subgraph on $(K-1)$ vertices. The precise details are left to the interested reader (who may wish to note exercise 8.13).

Thus KC remains NPC even if K is restricted to three. It is interesting to note that $3C$ remains NPC even with the further restriction that G is planar.

3-colouring Planar graphs (3CP)

Instance: A planar graph G.

Question: Does G have a (proper, vertex) colouring using three colours?

Before proving that $3CP$ is NPC we require the following lemma:

Lemma 8.2. In a 3-colouring of the following graph:

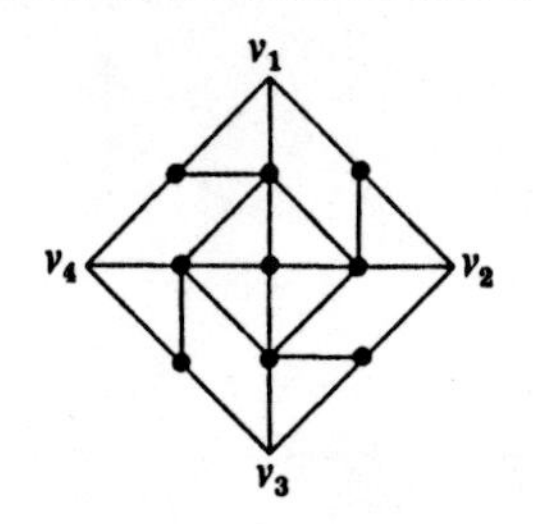

we have:

(*a*) v_1 and v_3 are identically coloured as are v_2 and v_4.
(*b*) v_1 and v_4 may or may not be similarly coloured.

Proof. As with lemma 8.1, we leave verification to the reader. It will be convenient to use the following short-hand for this planar graph in what follows. ■

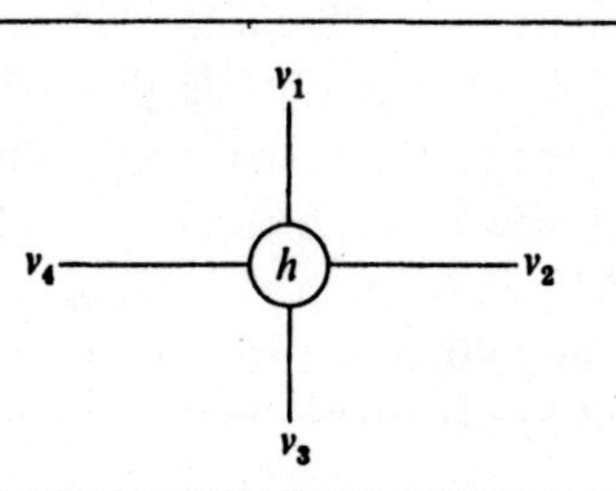

Theorem 8.11. $3CP$ is NPC.

Proof. We show that $3C \propto 3CP$. Let G' be an instance of $3C$, we construct an instance of $3CP$, G, as follows. G' is drawn on the plane in such a way that edges may cross, but not so that any edge touches a vertex other than its own end-points, also no more than two edges may cross at any one point. In general any edge (u, v) will be crossed by other edges as indicated in figure 8.8(a). We add new vertices along (u, v) one between each cross-over and one between each end-point and the nearest cross-over as shown

Fig. 8.8

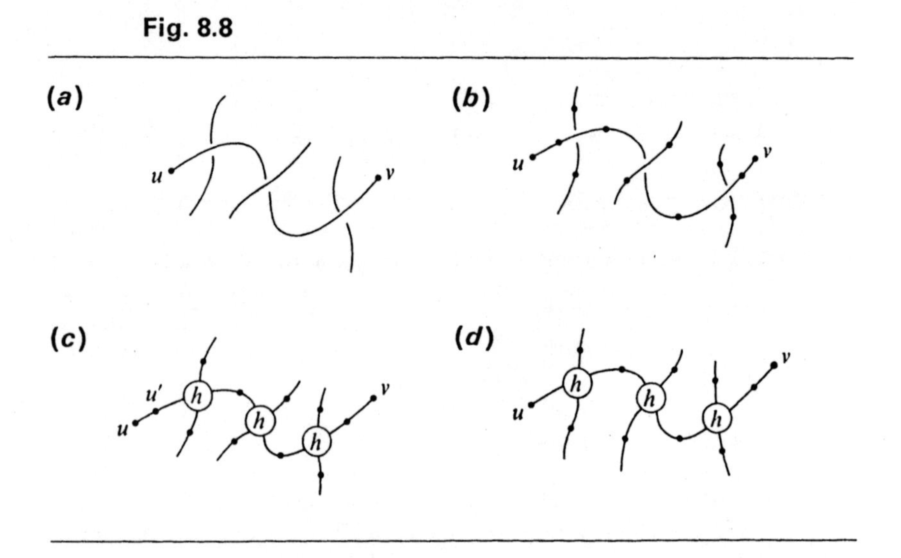

in (b). The next step is to replace each cross-over by a copy of the graph described in lemma 8.2. This is shown in (c). The result is clearly a planar graph. The final step in our construction of G is to choose *one* end-point of each original edge (u, v) and to contract the edge between that end-point and the nearest vertex along the old (u, v) edge. Thus (c) becomes (d), where (u, u') has been contracted.

Let V be the vertex-set of G and $V' \subset V$ be the vertex set of G'. It is easy to see that a 3-colouring of G is a 3-colouring when restricted to V' in G'. Conversely, suppose that G' has a 3-colouring. We can extend this to a 3-colouring of G as follows. For each original edge $(u, v) \in G$ with cross-overs, let u be the vertex which was coalesced with the nearest new vertex on (u, v). Colour every new vertex along (u, v) with the same colour as u. The interior vertices of each h-subgraph can then, according to lemma 8.2, be coloured using no more than our original three colours.

Thus $3C \propto 3CP$ and hence $3CP$ is NPC. ■

Finally, we show that the problem of finding the edge-chromatic index of a graph is *NPC*. In fact, we specifically prove the stronger result that it is *NP*-complete to determine whether or not the edge-chromatic index of a 3-regular graph is 3 or 4. Of course, the edge-chromatic index cannot be less than 3 and by Vizing's theorem (chapter 7) it cannot exceed 4.

Cubic graph edge-colouring (CGEC)

Instance: A 3-regular graph G.

Question: Does there exist a (proper, edge-)colouring of G using three colours?

Theorem 8.12. *CGEC* is *NPC*.

Proof. Clearly $CGEC \propto NP$. We complete the proof by outlining a transformation: $3SAT \propto CGEC$. In other words, we show how to construct, from an instance I of $3SAT$, a 3-regular graph G which is 3-colourable if and only if I is satisfiable.

G is constructed from a number of components each of which is designed to perform a specific task. These components are connected by pairs of edges such that, in a 3-edge-colouring of G, a pair represents the value *true* if both edges are identically coloured and if they are differently coloured then they represent the value *false*.

A key component of G is the *inverting component* shown in figure 8.9 along with its symbolic representation.

Fig. 8.9

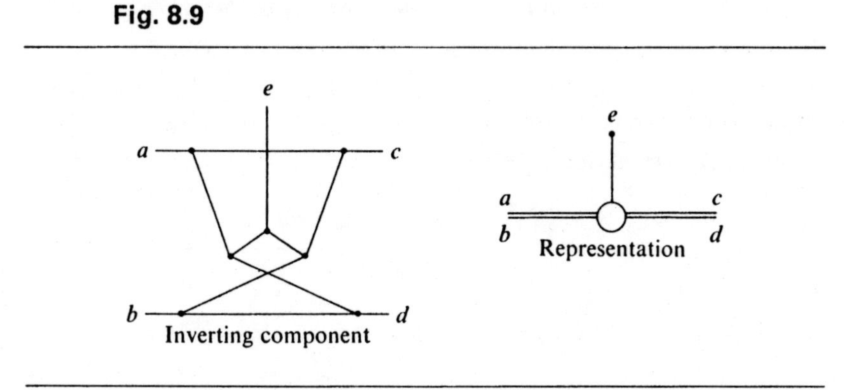

Lemma 8.3. In a 3-edge-colouring of an inverting component, the edges in one of the pairs of edges (a, b) or (c, d) are similarly coloured, whilst the remaining three labelled edges have distinct colours.

Proof. Like the previous two lemmas we leave this an exercise. However, the comment in exercise 8.9 may be of use. ■

If for the inverting component we look upon the pair of edges (a, b) as input and the pair (c, d) as output, then this component may be regarded as turning a representation of **true** into one of **false** and vice-versa.

We next describe a *variable-setting component* of G. Such a component exists for each variable v_i of the instance of $3SAT$. An example of such a component is shown in figure 8.10. This has four pairs of output edges. In general, however, there should be as many output pairs as there are appearances of v_i or its complement $\bar{v}_i$ in the clauses of the instance of $3SAT$. If there are x such appearances, then we can make, in an obvious way, a variable-setting component from $2x$ inverting components.

Fig. 8.10. A variable-setting component.

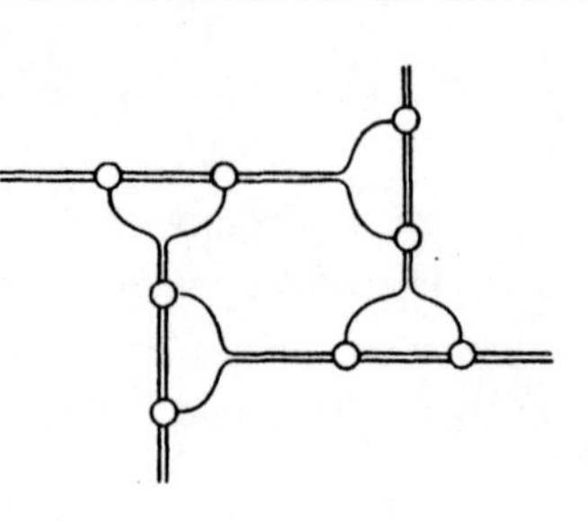

Lemma 8.4. In any 3-edge-colouring of a variable-setting component, all output pairs are forced to represent the same value, **true** or **false**.

Proof. Again, this is straightforward and is left as an exercise. ■

Finally we describe a *satisfaction-testing component* of G. Such a component is shown in figure 8.11. The required property of this component is embodied in the following lemma.

Fig. 8.11. A satisfaction-testing component.

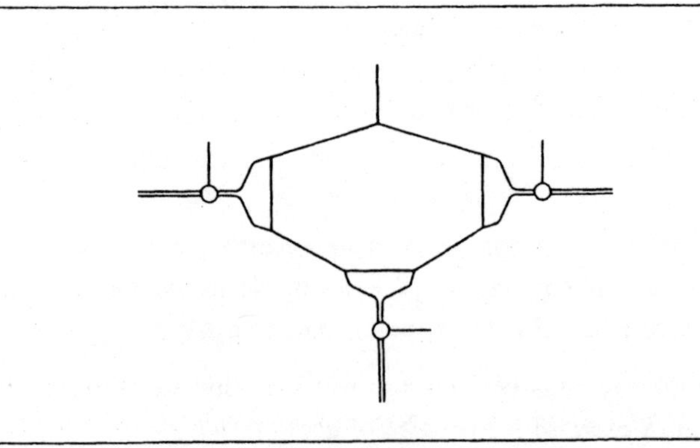

Lemma 8.5. A 3-edge-colouring of a satisfaction-testing component is possible if and only if at least one of the input pairs of edges represents the value **true**.

Proof. This is straightforward and is left as an exercise. ■

Given an instance I of $3SAT$ we now show how to construct the 3-regular graph G which is 3-colourable if and only if I is satisfiable. For each v_i, we construct a variable-setting component V_i which has an output pair of edges for each appearance of the variable v_i or its complement $\bar{v}_i$ amongst the clauses of I. For each clause c_j of I we have a satisfaction-testing component C_j. Let $l_{j,k}$ be the kth literal of c_j. If $l_{j,k}$ is the variable v_i then identify the kth input pair of C_j with one of the output pairs of V_i. Otherwise, if $l_{j,k}$ is $\bar{v}_i$ then place an inverting component between the kth input pair of C_j and the output pair of V_i. Let H be the graph resulting from this construction. H will have some unmatched connecting edges from the C_j. In order to construct the 3-regular graph G, we simply take two copies of H and join them together by identifying the unmatched edges.

It is easy to see that G can be constructed in polynomial time. We complete the proof by noting that the properties of the components, as described in lemmas 8.3, 8.4 and 8.5, ensure that G can be 3-edge-coloured if and only if the instance I of $3SAT$ is satisfiable. ■

8.3 Concluding comments

Figure 8.12 shows the tree of transformations we have developed through the theorems of this chapter which, along with Cook's theorem, establishes some members of the class of *NP*-complete problems. If $P \neq NP$ then, of course, no member of this class has an efficient algorithmic solution. On the other hand, if an efficient solution is found for one, then there exists an efficient algorithm for any other. A great deal of effort has been fruitlessly expended in the search for efficient algorithms so that there is widespread belief that $P \neq NP$. Graph theory contains a large number of problems that are *NP*-complete and the reader is referred to Garey & Johnson[1] for the complete list.

The number of problems in figure 8.12 is relatively small, although it includes perhaps the best-known *NP*-complete graph problems and certainly the most important regarding material in this book. In this respect we might also have included problems of multicommodity flow,[9] maximum cuts[10] in networks and the Steiner tree problem.[11]

The establishment of *NP*-completeness for a problem need not be the final point in consideration of its time-complexity. We can proceed in

Fig. 8.12

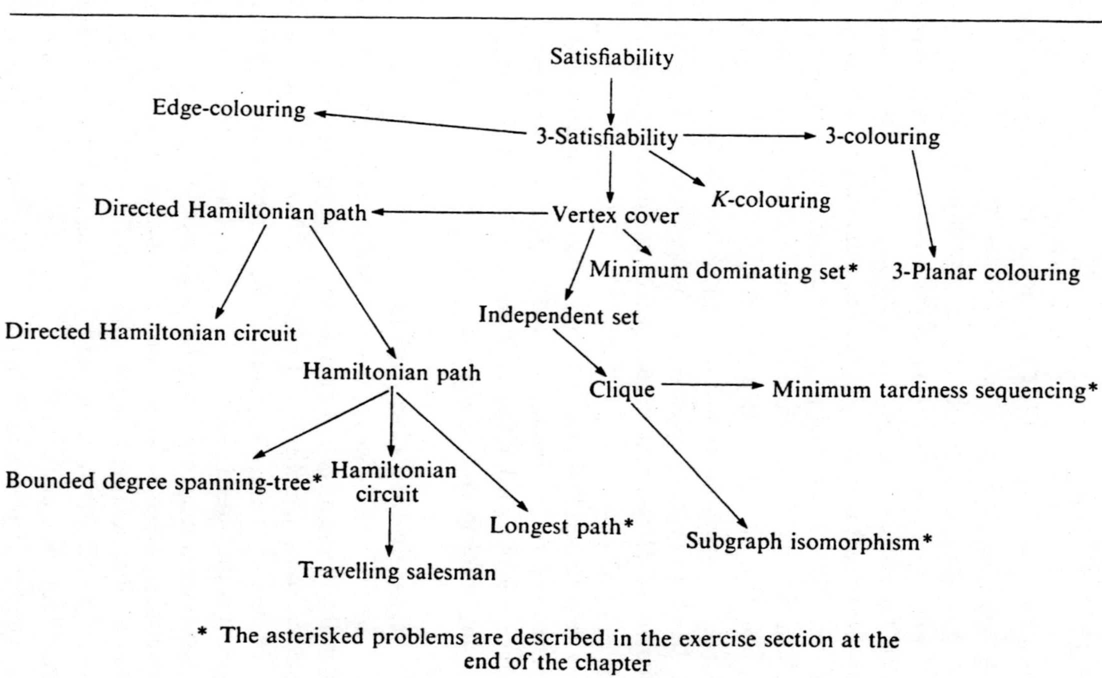

* The asterisked problems are described in the exercise section at the end of the chapter

several directions. First of all there might be a significant subclass of the problem that can be solved in polynomial time. For example, if we restrict the *NP*-complete problem of finding a maximum cut of a network to planar graphs then it becomes a member of the set *P*.[5] On the other hand, as we have seen, restricting the 3-colouring problem to planar graphs makes no difference to its *NP*-completeness. The point, however, is that such investigations may be fruitful.

A second line of inquiry that might be followed involves approximation algorithms: can we find an efficient (heuristic) algorithm which, within known bounds, provides an approximate solution to the problem? In chapter 6, for example, we approached the problem of finding feasible solutions to the travelling salesman problem in this way. It is an interesting fact that the theory of *NP*-completeness can be used to provide limits for the best possible approximations obtainable in this way. If A is the result acquired by an approximation algorithm (perhaps, for example, the length of a travelling salesman's tour) and if OPT denotes an exact solution, then (crudely) these limits are often expressed for all instances of the problem in hand, as a bound on the ratio A/OPT and sometimes as a bound on the difference $|A-OPT|$. A simple example of the latter is provided by exercise 8.15. Garey & Johnson[1] describe many results in this area.

Another approach to *NP*-complete problems which is of growing interest concerns *probabilistic analysis*. This involves, rather than concentrating on the worst-case behaviour of algorithms, the study of average-time performance or the evaluation of the exactness of approximation algorithms made under particular probabilistic assumptions. It can often be argued that knowledge of worst-case behaviour is of restricted value for practical purposes. It then makes sense to analyse the complexity and exactness as averaged over some distribution of instances of the problems. There will then be probabilistic guarantees for the results of such analysis. Slominski[13] provides a bibliography of such work.

The inefficiency of deterministic algorithms for *NP*-complete problems arises from (see exercise 8.6) the exponentially large number of solution 'guesses' that have to be handled sequentially. Whilst accepting an exponential-time complexity, it is nevertheless possible to minimise this expense by making a better definition of the objects to be searched than that implied by the entirely free description of an *NDTM* which was described earlier. Such an approach has been used for example for the problems of independent set[6] and *K*-colourability.[7]

8.4 Summary and references

In this chapter we have shown that a number of important graph problems belong to the large equivalence class of *NP*-complete problems which are widely believed to be intractable. Garey & Johnson[1] is an excellent general guide to the theory of *NP*-completeness while chapter 10 of Even,[2] which deals with graph problems, is also recommended reading. Aho, Hopcroft & Ullman[8] is a good general introduction to algorithmic design and complexity.

Cook laid the foundations for the theory of *NP*-completeness in a seminal paper[3] published in 1971. In that paper he stressed the importance of polynomial transformations, the class *NP* and proved that every problem in *NP* is polynomially transformable to satisfiability (*SAT*). Cook also showed that *CLIQUE* shared this property of being a hardest problem in *NP* and anticipated that many other problems would fall into the same category. Karp[4] shortly afterwards published such a collection of problems. These included 3-satisfiability and *K*-colourability. Since then a very large number of problems have been shown to be *NP*-complete, many of them in graph theory. Theorem 8.12 which establishes the *NP*-completeness of edge-colouring, is taken from Holyer.[12]

[1] Garey, M. R. & Johnson, D. S. *Computers and Intractability: A Guide to the Theory of NP-completeness*. Freeman (1979).
[2] Even, S. *Graph Algorithms*. Computer Science Press (1979).
[3] Cook, S. A. 'The complexity of theorem-proving procedures', *Proc. 3rd Ann. ACM Symp. on Theory of Computing*. Association of Computing Machinery, NY, 151–8 (1971).
[4] Karp, R. M. 'Reducibility among combinatorial problems', *in*: Miller & Thatcher (eds.), *Complexity of Computer Computations*, Plenum Press, NY, 85–103 (1972).
[5] Hadlock, F. O. 'Finding a maximum cut of a planar graph in polynomial time', *SIAM J. Comput.*, **4**, 221–25 (1975).
[6] Tarjan, R. E. & Trojanowski, A. E. 'Finding a maximum independent set', *SIAM J. Comput.*, **6**, 537–46 (1977).
[7] Lawler, E. L. 'A note on the complexity of the chromatic number problem', *Information Processing Letters*, **5**, 66–7 (1976).
[8] Aho, A. V., Hopcroft, J. E. & Ullman, J. D. *The Design and Analysis of Computer Algorithms*. Addison-Wesley (1974).
[9] Even, S., Itai, A. & Shamir, A. 'On the complexity of timetable and multicommodity flow problems', *SIAM. J. Comput.*, **5** (4) (1976).
[10] Garey, M. R., Johnson, D. S. & Stockmeyer, L. J. 'Some simplified *NP*-complete graph-problems', *Theor. Comput. Sci.*, **1**, 237–67 (1976).
[11] Garey, M. R., Graham, R. L. & Johnson, D. S. 'The complexity of computing Steiner minimal trees', *SIAM. J. Appl. Maths.*, **32**, 835–59 (1977).
[12] Holyer, I. 'The *NP*-completeness of edge-colouring', *SIAM. J. Comput.*, **10** (4) (November 1981).
[13] Slominski, L. 'Probabilistic analysis of combinatorial algorithms: A bibliography with selected annotations', *Computing*, **28**, 257–67 (1982).

EXERCISES

8.1. In order to solve a problem Q, we have one hour of computing time on a machine which operates at 2^{10} steps per second. Suppose that two algorithms are available for Q: A_1 of complexity n^6 and A_2 of complexity 2^n, where n is the problem size. Show that for any problem size, $n\ (> 1)$, A_2 is more efficient than A_1 for problems that can be solved within the available time.

What is the maximum problem size that can be handled by each algorithm within an hour? Above what value of n is A_1 more efficient than A_2?

(There are $\sim 2^{22}$ computational steps available in an hour.)

8.2. Show that the following problems (defined in the text) are in NP:
vertex-cover (VC)
directed Hamiltonian path (DHP)
3-colouring ($3C$)
and show that the transformations: $3SAT \propto VC$ (theorem 8.3), $VC \propto DHP$ (theorem 8.6) and $3SAT \propto 3C$ (theorem 8.10) are indeed polynomial.

8.3. Prove that MDS is NPC:
Minimum dominating set (MDS)
Instance: A graph G and an integer k.
Question: Does G contain a dominating set of size $\leqslant k$?
(It is easy to show that $VC \propto MDS$. Let G' and k be an instance of VC, then an instance of MDS consists of G (constructed from G' by adding, for every edge $e_i = (u, v) \in G'$, a new vertex x_i and edges (u, x_i), (x_i, v)) and k.)

8.4. Show (directly) that

$$CLIQUE \propto IS \propto VC$$

where the problems $CLIQUE$, IS and VC are defined in the text.

8.5. (*a*) Show that the DTM of figure 8.1 would verify that M divides N in $p(N)$ computational steps (i.e., movements of the tape head) where:

$$p(N) = \left(\frac{M+1}{M}\right) N^2 + \left(\frac{(M+2)^2}{M}\right) N - 1 = O(N^2)$$

(*b*) Redesign the quintuple set of figure 8.1 so that the same problem is solved but on an input tape which is the same as that shown in the diagram except that the tape cells containing A and B are blank. That is, remove A and B from T and model the original computation with the introduction of some new states and a modification of the quintuple set. This need not affect $p(N)$.

(*c*) Show that the $NDTM$ of figure 8.2 solves a problem that is in P.

8.6. Suppose that no efficient algorithm is known for the decision problem Q, but that $Q \in NP$. Show that an exponential-time deterministic algorithm exists for Q.

(Let the *NDTM* which verifies Q in time $p(n)$ be changed into a *DTM* which checks all possible guesses in series. The time required for this *DTM* is of order $p(n)|T|^{p(n)}$ where T is the set of tape symbols.)

8.7. Given a polynomial time algorithm to factorise an integer N, in unary representation, show that this *does not imply* that a polynomial time algorithm exists to factorise N in binary representation.
(Simply show that the length of N in a unary representation is exponential in terms of its length in binary representation.)

8.8. Let an arbitrary decision problem D, be: Given an instance I of the problem, is A **true** for I? The *complementary* decision problem, D^c, is then: Given I is A **false** for I?
Show that if D is a member of P then so is D^c. In contrast show that if D is a member of *NP* then we cannot necessarily draw the conclusion that $D^c \in NP$.
(If $D \in P$ then a *DTM* exists for D which halts within polynomial time for all I. D^c can then be solved by the same *DTM* by simply interchanging the states q_Y and q_N. On the other hand consider the case where D^c is the problem: Is it true that *there does not exist* a travelling salesman's tour of length $\leqslant k$?)

8.9. Justify the lemmas 8.1–8.5. For lemma 8.3, exercise 7.9 may be of assistance.

8.10. Show that the following problem is *NPC*:
Subgraph Isomorphism (*SI*)
Instance: Two graphs G_1 and G_2.
Question: Does G_1 contain a subgraph isomorphic to G_2?
(Consider the case that G_2 is a complete graph. In other words show that *CLIQUE* $\propto$ *SI*.)

8.11. Show that the following problem is *NPC*:
Bounded degree spanning-tree (*BDST*)
Instance: A graph G and an integer k.
Question: Does G contain a spanning-tree for which no vertex has degree $\geqslant k$?
(Consider the case $k = 2$. In other words show that *HP* $\propto$ *BDST*.)

8.12. In chapter 1 we saw that the problem of finding the shortest path between two vertices in a graph can be solved in polynomial time. Show, in contrast, that the following problem is *NPC*:
Longest Path (*LP*)
Instance: A graph G with two distinguished vertices u and v, and an integer k, $1 \leqslant k \leqslant n$.
Question: Does G contain a simple path from u to v of length $\geqslant k$?
(Show that *HP* $\propto$ *LP*.)

8.13. In analogy with lemma 8.1 let

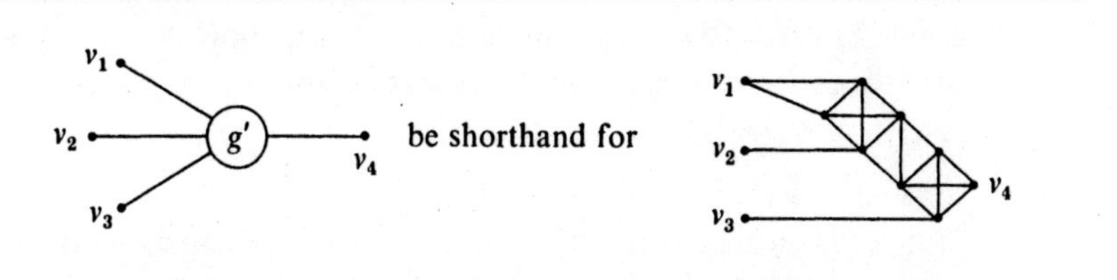

Justify the claim that in a 4-vertex-colouring of g' using the colour set $\{0, 1, 2, 3\}$, v_4 must be coloured 0 if and only if *all* of v_1, v_2 and v_3 are coloured 0. Use g' in a proof, similar to that for theorem 8.10, that $4C$ is *NPC*.

Notice that if we restrict $4C$ to planar graphs then, in view of the Appel–Haken proof of the four-colour conjecture, the problem is in *P*.

8.14. Consider the following problem:

Minimum Tardiness Sequencing (*MTS*)

Instance: A set of tasks $T = \{t_1, t_2, ..., t_p\}$ each requiring one unit of execution time, a set of deadlines $D = \{d(t_1), d(t_2), ..., d(t_p)\}$, a partial order $\lessdot$ on T and an integer K, $0 < K \leqslant |T|$.

Question: Is there a schedule (i.e., an order of execution of the t_i) such that if $t_i \lessdot t_j$ then t_i is executed before t_j and such that no more than K tasks are completed after their deadlines?

Given an instance of *CLIQUE*, *I*, we can construct an instance $F(I)$ of *MTS*. Let I consist of $G = (V, E)$ and an integer L. $F(I)$ then consists of:

$T = \{v_1, v_2, ..., v_m, e_1, e_2, ..., e_l\}$, $v_i \in V$ and $e_j \in E$

$v_i \lessdot v_j$ if in G, v_i is an end-point of e_j

$d(v_i) = \frac{1}{2}L(L+1)$, $d(e_j) = |V|+|E|$

$K = |E|-(\frac{1}{2}L(L-1))$

This construction can clearly be carried out in polynomial time. Show that I has a clique of size L if and only if $F(I)$ has no more than K tasks completed after their deadlines. This will prove that *MTS* is *NPC*. (The partial order requires that in every schedule any 'edge' task is completed after its own 'end-point' tasks. The deadlines are such that only 'edge' tasks can be late. If the answer to *MTS* for $F(I)$ is 'yes', then at least $\frac{1}{2}L(L-2)$ of these tasks must be completed before their deadline, $\frac{1}{2}L(L+2)$. In order that the schedule does not violate the partial ordering, the corresponding 'vertex' tasks must also be executed before this deadline. The *minimum possible* number of these is L (when the corresponding vertices in G form a clique) so that the total number of tasks now performed before the time $\frac{1}{2}L(L+1)$ is

$\frac{1}{2}L(L-1)+L = \frac{1}{2}L(L+1)$

This just exhausts the available time before the 'edge'-task deadline.)

8.15. Let $A(G)$ denote the number of vertices that an arbitrary approximation algorithm A assigns to a maximum independent set for the graph G. We denote by $OPT(G)$ the exact number in this set. Show that if $NP \neq P$, then the proposition that for all instances G that:

$$|A(G) - OPT(G)| \leqslant K$$

for some integer constant K, is false.
(Suppose that the proposition is true. Given G we can apply A to the (disconnected) graph G' which consists of $(K+1)$ copies of G. Then

$$|A(G') - OPT(G')| \leqslant K$$

and clearly $OPT(G') = (K+1)\ OPT(G')$. This also defines an algorithm B which assigns at least $\lceil A(G')/(K+1)\rceil$ vertices to the maximum independent set of G – obtained by finding the largest set of vertices that A assigns to any of the components of G'. It follows that:

$$|B(G) - OPT(G)| \leqslant K/(K+1)$$

In other words, $B(G)$ must be equal to $OPT(G)$ so that B would be an exact algorithm. With theorem 8.4 this provides a contradiction.)

Appendix

On linear programming

Several problems in this text can be formulated as *linear programming problems*. For example, the *minimum-cost flow algorithm* of chapter 4 was described in this way. We present here just enough insight into linear programming theory for an understanding of its application within this text. Readers who require an extensive treatment should consult one of the numerous texts ([1], for example) devoted to the subject.

A linear programming problem is any problem that can be described as follows:

$$\text{maximise } \sum_{i=1}^{n} c_i x_i \qquad \text{(i)}$$

subject to the constraints:

$$\sum_{i=1}^{n} a_{ji} x_i = b_j,\ 1 \leqslant j \leqslant k$$

$$\sum_{i=1}^{n} a_{ji} x_i \leqslant b_j,\ k+1 \leqslant j \leqslant m \qquad \text{(ii)}$$

and the non-negativity conditions:

$$x_i \geqslant 0,\ l+1 \leqslant i \leqslant n \qquad \text{(iii)}$$

where for $1 \leqslant i \leqslant l$, x_i is unrestricted in sign. Within (i), (ii) and (iii) the a_{ji}, b_j and c_i are given constants while the x_i are variables.

The above definition of a linear programming problem is in one of a number of standardised forms in common use. Many linear programming problems have a natural description in statements similar to, but not immediately identical to (i), (ii) and (iii). Fairly trivial adjustments can cast such problems into our standardised form. For example, a problem of *minimising* $\sum_i c_i x_i$ is the same as *maximising* $\sum_i (-c_i) x_i$, a linear constraint of the form $\sum_i a_{ji} x_i \leqslant b_j$ is the same as $\sum_i (-a_{ji}) x_i \geqslant (-b_j)$, and so on.

In order to provide motivation for the above abstractions we briefly describe one of the problems (first published in [2]) which led to the development of linear programming theory. This problem, known as the *dietician's problem*, is to determine a week's diet for a hospitalised patient who needs a predetermined minimum weight of each of the nutrients $N_1, N_2, \ldots, N_m$. There are n different foods available $F_1, F_2, \ldots, F_n$ and the nutritional content of each is known. The difficulty is that the cost of the diet has to be minimised. With the following definitions:

a_{ij} is the number of units of N_i in one unit of F_j
b_j is the minimum number of units of N_j to be consumed
c_i is the unit cost of F_i

The dietician's problem can then be simply stated as follows:

$$\text{minimise} \sum_{i=1}^{n} c_i x_i \qquad (a)$$

where x_i is the number of units of F_i consumed, and the weekly consumption of N_j must be at least b_j units:

$$\sum_{i=1}^{n} a_{ji} x_i \geqslant b_j,\ 1 \leqslant j \leqslant m \qquad (b)$$

while the consumed amount of each food must be non-negative:

$$x_i \geqslant 0,\ 1 \leqslant i \leqslant n \qquad (c)$$

It is an easy matter to convert (*a*), (*b*) and (*c*) into our standard form by the means indicated earlier. What is obtained has $k = 0$ for (ii) so that all the constraints are inequalities, and has $l = 0$ for (iii) so that every variable is restricted in sign.

In general a linear programming problem has n variables x_i, an *objective function* (i), m *constraints* (ii) and a set of *non-negativity conditions* (iii). For any such problem we can construct a *dual* problem with m variables y_i (one corresponding to each constraint of the original problem) and n constraints (one for each variable of the original problem) as follows:

$$\text{minimise} \sum_{j=1}^{m} b_j y_j \qquad (\text{i}')$$

subject to the constraints:

$$\left.\begin{aligned} &\sum_{j=1}^{m} a_{ji} y_j = c_i,\ 1 \leqslant i \leqslant l \\ &\sum_{j=1}^{m} a_{ji} y_j \geqslant c_i,\ l+1 \leqslant 1 \leqslant n \end{aligned}\right\} \qquad (\text{ii}')$$

and non-negativity conditions:

$$y_j \geqslant 0,\ k+1 \leqslant j \leqslant m \qquad (\text{iii})$$

while for $1 \leqslant j \leqslant k$, y_j is unrestricted in sign. Notice that the coefficients a_{ji} in the constraining relations are the transpose of those in (ii). The coefficients a_{ji}, b_j, c_i, l and k are fixed as in (i), (ii) and (iii). The dual problem can easily be written in the standard form (i), (ii) and (iii) and so is itself a linear programming problem. In order to distinguish it from the dual, the original problem is called the *primal.* If the dual is written in the form of the primal and its dual constructed according to the above recipe, then it is easy to see that the dual of the dual is the primal.

Given the meaning of the parameters a_{ij}, b_i and c_i afforded by a specific primal problem, useful insight can be gained from an interpretation of its dual. The dual of the dietician's problem (*a*), (*b*) and (*c*) is:

$$\text{maximise} \sum_{j=1}^{m} b_j y_j \qquad (a')$$

subject to:

$$\sum_{j=1}^{m} a_{ji} y_j \leqslant c_i, \ 1 \leqslant i \leqslant n \qquad (b')$$

and

$$y_j \geqslant 0, \ 1 \leqslant j \leqslant m \qquad (c')$$

which can be interpreted as follows. Knowing of the dietician's problem a chemist plans to manufacture pills which can be used as a substitute for food in a diet. The chemist can only hope to sell his pills if they provide a cheaper alternative to the food, but at the same time he must maximise his profit. If y_i is the price of one unit of the N_i pills, then the price of the pill equivalent of a week's minimum nutritional intake, which has to be maximised, is precisely expressed by (a'). Moreover (b') is the requirement that the pills be cheaper than the food and (c') states that the prices of the pills must not be negative.

A set of values for the variables of a linear programming problem which satisfies the constraints and non-negativity conditions is called a *feasible solution.* If a feasible solution optimises the objective function then it is called an *optimal solution.* The value of the objective function obtained with an optimal solution is called the *value of the linear program.*

Given a feasible solution to a linear programming problem, we need to know whether or not it is an optimal solution. We shall now see how this might be ascertained. If the sets of values $\{x_i\}$ and $\{y_i\}$ are, respectively, feasible solutions for a primal and its dual then from (ii) and (ii′) we see that:

$$\sum_{j=1}^{m} b_j y_j \geqslant \sum_{j=1}^{m} y_j \sum_{i=1}^{n} a_{ji} x_i = \sum_{i=1}^{n} x_i \sum_{j=1}^{m} a_{ji} y_i \geqslant \sum_{i=1}^{n} x_i c_i \qquad \text{(iv)}$$

where the objective function for the primal is on the right and for the dual is on the left. From (iv) we see that the value of a maximisation problem cannot exceed the value of its dual. For the dietician's problem this means that the diet in pill form cannot be more expensive than if the food is used. If feasible solutions can be found for both the primal and the dual such that equality holds throughout (iv), then, clearly, optimal solutions have been found. This is the case if and only if

$$\left(b_j - \sum_{i=1}^{n} a_{ji}x_i\right) y_j = 0,\ 1 \leqslant j \leqslant m \tag{v}$$

and

$$\left(\sum_{j=1}^{m} a_{ji}y_i - c_i\right) x_i = 0,\ 1 \leqslant i \leqslant n \tag{vi}$$

For problems of interest to us we obtain solutions which satisfy (v) and (vi) and which must therefore be optimal. The important relations (v) and (vi) are often expressed in the following form:

$$y_j \neq 0 \Rightarrow b_j = \sum_{i=1}^{n} a_{ji}x_i,\ 1 \leqslant j \leqslant m$$

and

$$x_i \neq 0 \Rightarrow c_i = \sum_{j=1}^{m} a_{ji}y_i,\ 1 \leqslant i \leqslant n$$

and are known as the *complementary slackness conditions.*

The standard method of solving linear programming problems is called the *simplex method* (see [1], for example). Starting from one feasible solution this method generates a sequence of others such that each subsequent solution produces a better value for the objective function. Eventually an optimal solution is produced. Where we need to apply the theory in this text, we describe specific means for generating a sequence of feasible solutions and we show that this terminates with an optimal solution.

Linear Programming is a good example of a problem which is more effectively solved by an exponential-time algorithm (in this case the simplex method) than by a known polynomial time algorithm (in this case the ellipsoid method). Klee & Minty[3] have shown that the simplex method is exponential-time in worst-case performance, yet in practice it has, as McCall[4] shows, exhibited linear-time behaviour. Smale's[5] goal is to provide a theoretical explanation. The ellipsoid method is due to Khachian.[6]

[1] Trustrum, K. *Linear Programming*, Routledge and Keegan Paul (London) (1971).

[2] Steigler, G. J. 'The cost of subsistence', *J. of Farm Econ.*, **27**, 303–14 (1945).

[3] Klee, V. & Minty, G. J. 'How good is the simplex algorithm?', *in*: O. Shisha (ed.), *Inequalities* III, Academic Press, NY, 159–75 (1972).
[4] McCall, E. H. 'Performance results of the simplex algorithm for a set of real-world linear programming models', *CACM*, **25** (3), 207–13 (1982).
[5] Smale, S. 'On the average number of steps of the simplex method of linear programming', *Math. Program.* (Holland), **27** (3), 241–62 (1983).
[6] Khachian, L. G. 'A polynomial time algorithm for linear programming', *Doklady Akad. Navk SSSR*, **245**, 5, 1093–6 (1979). Translated in *Soviet Math. Doklady*, **20**, 191–4.

Author index

Subject index

9 780521 288811